MW01259289

Mastering the Fire Service Assessment Center

Mastering the Fire Service Assessment Center

Anthony Kastros

Disclaimer

The recommendations, advice, descriptions, and the methods in this book are presented solely for educational purposes. The author and publisher assume no liability whatsoever for any loss or damage that results from the use of any of the material in this book. Use of the material in this book is solely at the risk of the user.

PennWell Corporation
1421 South Sheridan Road
Tulsa, Oklahoma 74112-6600 USA

800.752.9764
+1.918.831.9421
sales@pennwell.com
www.pennwellbooks.com
www.pennwell.com
www.FireEngineeringBooks.com

Director: Mary McGee
Managing Editor: Jerry Naylis
Production/Operations Manager: Traci Huntsman
Production Editor: Tony Quinn
Cover Designer: Matt Berkenbile
Book Designer: Sheila Brock

Library of Congress Cataloging-in-Publication Data

Kastros, Anthony.

Mastering the fire service assessment center / by Anthony Kastros.

p. cm.
ISBN-13: 978-1-59370-077-5 (hardcover)
1. Fire extinction--Vocational guidance. 2. Fire extinction--Examinations--Study guides. 3. Firefighters--Job descriptions--Handbooks, manuals, etc. 4. Firefighters--Selection and appointment--Handbooks, manuals, etc. 5. Applications for positions--Handbooks, manuals, etc. I. Title.

TH9119.K37 2006

363.37023--dc22

2006018093

Printed in the United States of America

3 4 5 6 11 10 09 08

To my wife Cynthia and daughter Sophia.

You are my true companions.

All my love, all my life.

Table of Contents

Foreword

Battalion Chief Anthony Kastros asked me to write a foreword for his book on fire service assessment centers. I read, enjoyed, and learned a lot reading his book. The book is an excellent presentation that describes how to prepare for an extremely important personal and professional process.

The entry and promotional process is typically a stressful, emotional, and very high profile fire department event. It is the most competitive organizational process that formally occurs in our professional lives. Assessment centers create the foundation for an accepted and legitimate way for the organization to rearrange the regular day-to-day, continuous and highly observed *line of humans* that structures and staffs our departments. It provides an opportunity for qualified members to apply and present themselves to the organization, competing for a new spot to first get them into the organization (entry) and then to move up inside the organization (promotion). It creates a standard internal structure that provides a steady stream of upwardly mobile candidates that are necessary to staff all of the positions on every level within our departments.

On a very personal level, it is a big deal in a fire department where you get to sit. Our seat reflects our position and our status. When you first arrive, you get to sit in a seat that is facing backward. Your first move then is to get to sit behind the steering wheel. A big change is to then get to move into the company officer seat. This is the first boss seat (right front seat in the rig/head of the table in the station). Then you get to change vehicles and sit in the battalion chief (BC) buggy. If you are nuts (to stop being a BC) and you keep showing up to the assessment centers, you get to go "downtown" to the "puzzle palace" and go through a series of boss seats in the administrative part of the department.

Finally, they have some cake and punch, give you a gold watch, and you get to go sit on the front porch of the old firefighters' home and watch the fire trucks go by.

The way you get to personally change seats is to participate in the formal organizational process where you present yourself to a group of "forensic" interviewers and they do an "occupational autopsy" on you. As you go through the "procedure," they get to remove, examine, and evaluate all your pieces and parts. Then they make a judgment if those components have the capability and potential to be rearranged to somehow become the position you are competing for. Necessarily, trying to somehow get ready for your "autopsy" is pretty nerve-wracking—simply, it's scary to think the board may disassemble you and then put you back together and send you back out to be just what you were before.

Here's where Anthony comes in. He has developed and packaged a simple, straightforward, understandable, nine-step preparation that describes everything in nice plain firefighter language. His process shows how to get ready for, approach, and most of all, effectively pull off hopping up on the stainless steel table and actually surviving—and perhaps even enjoying the experience.

Chief Kastros is a member of the current class of exceptional *young Turk* fire officers. He is youthful, smart, experienced and thoughtful. As you read his book, you will be persuaded to be your best by his positive approach and attitude. His writing reveals that he has thought about, planned for, and gone through the process he writes about. He writes (thankfully) about what it takes to both *get* the job and what it takes to *do* the job. The get-the-job and do-the-job parts must be closely connected if the assessment center process is valid. The basic reason we go through any organizational effort (including the assessment center) is to create and maintain the organizational ability to effectively deliver teams of workers to Mrs. Smith when she needs us. It is critical that every member can play their team role. How they are selected is critical to that team success.

I have known Anthony a long time, and his words describe how he feels and what he does. My only regret about him is that his parents did not have car trouble in Phoenix on their way to Sacramento because if they had, he would now be a battalion chief on the Phoenix Fire Department instead of the excellent department where he now resides.

Pay attention as you read his book because I now hear the forensic team sharpening their instruments—you're next. They are ready for you so you better be ready for them.

Alan V. Brunacini
Fire Chief, Phoenix Fire Department

Acknowledgments

First, I must thank God. Without the flood of blessings, people, inspiration, and answered prayers, this book would not be conceived of, let alone possible.

Thank you to my brothers, Mitch and Deme Kastros. You have paved the way in this business and inspired me beyond words. Thank you both for being my best friends, brothers in life, and brothers in battle.

Thank you to Fire Chief Alan V. Brunacini for his inspiration, influence, mentoring, wit, humor, support, and example. Without you, this book would never be dreamed of in the first place. In addition, thank you to Ellen and Kathi. You have a pretty great boss, but he is twice as blessed to have you both.

Thank you to Jerry Naylis from *Fire Engineering*/Pennwell. Your amazing support, professionalism, guidance, and tireless effort made this possible. We spoke so much on the phone that I started to talk with an East Coast accent.

Thank you to Fire Chief Don Mette and the Sacramento Metropolitan Fire District for your support and faith in the project. Thank you to Cristy Haverty for the pictures, your unyielding standards and energy, and the support for so many other things we do. You make it so much fun. Thank you to Assistant Chief Dave Baltzell and the folks at Metro Fire's Training Division for jumping up to help without hesitation: Metz, Wiggy, Coach, Tommy, and Randy.

Thanks also to Dan Haverty for your personal and professional brotherhood and kinship. Thanks to the gang at Station 114 for stepping in last minute. Thank you to Division Chief Stewart Roth for his awesome example, boundless enthusiasm, and sparkle in the eye.

Thanks to the USAF Thunderbirds. You had nothing to do with this book; you just helped shape the author who wrote it. Danny will live in you forever. Thanks ma! *S'Agapo.*

Writing a book has always been a dream for me. Dreams are not worthwhile unless you share them with someone most special. Finally, I thank my amazing wife Cynthia for her tireless support, input, readings, prayer, and belief. You make the journey worth it, no matter the destination. You are the greatest wife and best friend any man could hope for. I love you.

Introduction

So, you want to promote. We hope this book will help you attain your goal. Each year, thousands of firefighters take promotional assessment centers for fire officer ranks. Most of these folks don't understand the amount of preparation that is required for the *position* they seek. They only focus on the *test*. Aspiring officers who only focus on the test—without taking into consideration the position—are putting the cart before the horse. They eventually find themselves taking tests over and over without understanding why. Taking more tests without positional preparation will not yield miraculous changes, only additional frustration. You must know where you need to improve, develop a plan, and implement it.

Most firefighters do not experience an assessment center until they attempt promotion, although a few fire departments use them now for entry-level hiring. Traditional testing, both in college and the fire service, conditions us to study for written exams, oral interviews, and prepare for physical agility tests. Consequently, candidates for promotion spend the majority of their time studying for the written test because they are comfortable with that type of evaluation. When it comes to the assessment center, most are confused about how to prepare.

The goal of this book is to help you master the fire service assessment center; however, you must do more than read the contents of this text. You must also practice and apply the principles taught well beyond the time it will take you to finish this book. This book will teach you what the knowledge, skills, and abilities (KSAs) are to be an excellent officer, and how assessment centers utilize these KSAs to evaluate behavior, performance, and overall readiness for the job.

Notice that we said, "Teach you what the KSAs *are*." We didn't say, "Teach you the KSAs." That's the whole point. *You cannot change behavior from just reading a book. It's impossible. Instead, you will learn what you will need to learn, and how to learn it.* Make sense?

Think of the *Firefighter Combat Challenge* as an example. In just a few minutes, you could learn what the events are, what the records are, and that you need cardiovascular stamina, strength, endurance, and conditioning to be successful. Just knowing about those KSAs doesn't make you a champion Combat Challenge superhero. Given that knowledge, you *then* need to spend months, even years *practicing and developing* those KSAs (proper diet, exercise, and training) to have championship *performance*. The same principle applies to becoming a great officer *and* testing like one.

Assessment centers are about performance, behavior, and ability. You would not *study* for a physical agility test; you would *practice* (see fig. 0–1). The same applies to an assessment center.

You should not limit your preparation for the assessment center to studying. Although studying your policies, standing operating procedures (SOPs), and other components of the job may comprise part of your preparation, you must broaden the scope of your overall plan. *Developing and practicing the KSAs are the keys to success.*

You will be instructed how to formulate a plan for career success *both* in the testing process (assessment center), and more importantly, the position (promotion).

Fig. 0–1 Just studying for an assessment center makes as much sense as this picture. Is this guy going to get in shape by just studying this book?

This book is *first* about becoming an excellent officer, and then how that preparation for the position will help you *get* the position through the assessment center process. Make no mistake about it, you will learn more about assessment centers, the exercises used, and how to develop successful performance. The key is to keep the "horses" of good KSAs in front of the "cart" of an assessment center. There are no shortcuts.

This book will help you regardless of the fire officer rank you seek (lieutenant, captain, battalion chief, division chief, etc.). Assessment center exercises tend to remain consistent, regardless of rank, but the complexity tends to increase as you rise up the chain of command. We will point out specifics as we go.

Now we're past "So what do you want to get out of this book?" Let's now talk about what you *had better* get out of this book. You will not have gotten your money or time's worth from this book if you don't learn how to bridge the *gap* between *where you are* and *where you want to be.* If you want to finish this book knowing where you need to improve, how to develop a specific/personal plan to become an excellent officer, *and* do well in whatever assessment center exercises they throw at you as a result, then read on.

This book is comprised of nine main sections, all of which are critical to the successful performance in an assessment center: 1) elimination of barriers, 2) assessment center orientation, 3) mentality, 4) KSAs, 5) exercise-specific tools and key points, 6) sample exercises, 7) general key points and common pitfalls, 8) developing your plan, and 9) interviews.

Terminology

This book is designed to be usable by anyone facing a fire department assessment center, regardless of where you live in the United States. That being said, there are always differences in regional terminology in this business. For example, we use the incident command system (ICS) in this book. Many other forms exist in the country (incident management system, or IMS, for example). Just place your regional command system in the appropriate place. Also, the terms *SOP* and *standard operating guideline (SOG)* are synonymous for the purpose of this book. Whatever your department uses should be your term of choice.

How to Get the Most out of this Book

This is not designed to be a *read once, put on the shelf, collect dust, and burn when you get promoted* text. This is a workbook. You should plan to read it more than once. Some sections you may read three or four times. If you are like me, you'll read some 10 times before it sinks in. The point is, highlight, underline, rip, mark, and do whatever you must to get the most out of this book. There's plenty of space on the pages for a reason. Write whatever comes to mind as you go.

This book is compartmentalized. That's a fancy way of saying you can section it off. Chapter 6, for example, is nothing but mock exercises for you to copy, perform, re-take, and use as coasters if you like. Although the book has nine chapters that each function relatively independently, you must read the whole thing to get the most out of this book.

The reason you don't get rid of it when you promote is because you may see some of this again when you move up the chain of command and take more assessment centers. I have used all of the principles in this book to promote to captain and battalion chief. These principles have worked for countless firefighters that we've coached over the years. Quite simply, it works. Lots of #1 assessment center scores have come from this curriculum (including three of my own), and many more top 10 scores.

Here's the catch: It won't work unless you practice, train, and repeat. Reading this once may give you some tools, but you will not change your behavior unless you take on a long-term plan. *Remember, at the moment of truth—whether in a test or during the real thing—you will not rise to the level of expectation; you will fall to the level of training.*

Abbreviations

BC – battalion chief
CAN – conditions, actions, and needs
EAP – employee assistance program
EMS – emergency medical services
EMT – emergency medical technician
FATS – fire ground accountability tracking systems
FDIC – Fire Department Instructors Conference
FF – firefighter
FIRESCOPE – Firefighting Resources of California Organized for Potential Emergencies
FPODP – facts, probabilities, own situation, decision, plan of operation
HAZMAT – hazardous materials
HOT – hands-on training
IAP – incident action plan
IC – incident commander
ICP – incident command post
ICS – incident command system
IMS – incident management system
KSAs – knowledge, skills, and abilities
LCES – lookouts, communications, escape routes, safety zones
MCI – multi-casualty incident
MBO – management by objectives
OCAA – object, conditions, actions, assignments
ops – operations
OSHA – Occupational Safety and Health Administration
PAR – personnel accountability report
PIO – public information officer
PPE – personal protective equipment
probie/probies – probationary firefighter
RBO – relationship by objectives
RECEOVS – rescue, exposures, confinement, extinguishment, overhaul, ventilation, salvage
RIC – rapid intervention crew
RP –reporting party
SEE – safe, effective, and efficient
SOP – standard operating procedure
SOG – standard operating guidelines
SWOT – strengths, weaknesses, opportunities, and threats
USAR – urban search and rescue
WMD – weapons of mass destruction

Elimination of Barriers 1

Before we delve too far into assessment center specifics, we must first eliminate some barriers to success. These barriers often affect our performance in the assessment center without our even realizing it. We may be destined for poor performance before we even sign up for the test. By eliminating these barriers early, you can be free to reach your potential, focus on the KSAs, and practice the exercises. The barriers we will attack are fear (of failure), nerves, attitude, and baggage.

Do I Really Want This? What if I Fail?

Let's face it. The thought of taking a promotional test of any kind does not fill the typical firefighter with joy and elation.

Some of us would rather take a ball-peen hammer to the temple than study for months in preparation for a test. Most of us would rather go hunting, fishing, golfing, or play paint ball with our kids. The house needs work, our significant other wants quality time, and the kids are growing at an astronomical rate. You may be a soccer mom or dad, you may have a side business (imagine that, a firefighter working a second job), or you may just not really feel like you can give it your all at this time in your life (see fig. 1–1).

Fig. 1–1 It's easy to spend time wondering whether the promotional process is worth the time and effort.

For averaging some 20 days off a month, we sure can find a million ways to fill the time, and many of them important and worthwhile. Perhaps you are instructing classes, are a member of a rescue team, or are building your own house (all of which would help you be a better officer, by the way). The moral of the story is that a lot of things are competing with your potential even before you compete with the other candidates or fill out the application form for the test. Ever wonder how the rest of the world that works a nine-to-five job can get the lawn mowed, pay bills, and still have timc left for a ball game?

One key point is to sit down with your significant others and discuss the commitment required to be successful in promotion. The time you spend is worth it, but only if your family agrees. Free yourself from conflict by discussing the time you need, the family commitments you have, and get on the same page with your loved ones. If you pre-agree to commit to the process, then you will free yourself to stay focused.

Most of us are riddled with mixed feelings, nerves, frustration, and doubt when it comes to promotional testing. At least I was. Mixed feelings come from a myriad of sources: wondering whether it's worth the effort or if we have what it takes, fearing failure, not wanting to let down our families, or just plain not knowing if we want the job in the first place.

After all, "I've got a lot of seniority right in this spot. I could retire as a firefighter." Sound familiar?

How about this one, "The engineer is the greatest job in the fire service. I'm kind of my own boss, and the captain doesn't mess with me that much."

This one's my favorite: "I love my crew. If I promote, who knows where I will work or if I will like my new crew?"

If any of these sound familiar, then you are one of many who start rationalizing the test and the position away. *Even if you still take the test, you have given yourself a rational, self-bargained way out of succeeding. This makes failure an option and more palatable. Whenever failure is an easier option, it's more likely to happen.*

Who would you rather follow into battle? The leader who says, "Failure is not an option" or the one who says, "Hey, we might not make it." Even if the first guy is wrong, his team has a better chance of success going onto the battlefield.

Why do we hate failure? Because it's no fun. Let's admit it without any psychobabble. Society and especially the fire service society hate failure. We do not accept lost lives, property, wages, or football games that well. The typical firefighter is a competitive individual. We compete for station bids, seniority, combat challenges, response times, first-due fires, and behind-the-station basketball games. Heck, we even have our own Olympics for all of the competitiveness to find an outlet. Just look at the brochure to the Firefighter Olympics. There's everything from A to Z: albatross tossing to zebra tipping. We are a weird group of people.

Failing in a promotional test can feel like you have failed as a firefighter—if you let it. We may think our peers look at us differently. The fact is they don't. How many times have you heard an awesome firefighter or engineer described like this? "He'd be an awesome officer, but he just isn't a good test taker." If you fail a promotional test, congratulations! At least you tried, and now you have nothing to lose on the next one. Chances are your comrades have come to your aid, like the fallen firefighter. They have dusted you off and told you that you are a great member of the crew, a fantastic firefighter and will indeed make a great officer some day. That's the beautiful thing about us. We are competitive as can be, until someone gets hurt. Then we are there for each other, without hesitation.

Overcoming Fear of Failure and Nerves

You can overcome your fear of failure and nerves more easily than you think. Let's start with failure. First, recognize that failure is not the problem. Fear is the problem. Fear paralyzes and causes doubt. Remember FDR's classic speech, "The only thing we have to fear is fear itself." Once you identify the true problem, you can solve it.

Second, remember that you are the only one who can let fear in and keep it out. In other words, you can control it. Fear is an enemy that infiltrates your mind, *but only if you let it*. Choose not to be afraid. Do not underestimate the power of positive thinking. The human mind has put us on the moon, discovered electricity, nuclear energy, and the Halligan. Just think what your mind can do to keep you positive. No one can do that for you. It's up to you to choose. Remember that you cannot always control an outcome, but you *can* control fear.

Third, ask yourself, what am I really afraid of? What is causing the fear? Is it a damaged reputation, letting your family down, or the exposing of a weakness? As mentioned previously, your family and friends are going to be more proud of you for trying than for the outcome, and you will have grown whether you get promoted or not. At the very least, you will be closer to your ultimate goal.

Don't create fear where it doesn't exist. Eliminate it like an enemy that's trying to attack. If you think you will let your family down, ask them. If you think your crew will think less of you, ask them (see fig. 1–2). Chances are both groups will be much more supportive than you give them credit for, and you are probably creating unnecessary fear. Once you hear that they are proud of you for trying, you can remove the fear.

Fourth, you must realize that the greatest things in life are sometimes accomplished differently or later than we want. When I graduated high school, I wanted to go to college at Cal Poly in San Luis Obispo. When I was denied admission, I thought the world ended. Looking back now, that denial letter was the greatest thing that could have happened in my life at the time. If I had been accepted, I would not be blessed with my beautiful wife and daughter or the great place I work and live. Wouldn't it be nice to have that kind of foresight? You never know where the road of life will lead.

Another example was when I was denied the right to take a promotional test for captain even though I exceeded the minimum requirements. The person in charge of the test didn't think that my 60-unit Associate Degree in Fire Science met the required 30-unit fire certificate. My appeal was denied. I had no control over the outcome and felt very helpless as I watched everyone else take the test that I studied two years to take. Two years later, when I received the highest score on the captain test, my peers were much more supportive than before because they felt I had earned it. Looking back, I would not have changed a thing. Having the respect of our peers is much more important than a rank.

Fig. 1–2 We can fear losing the respect of our peers if we do not score well. Eliminate the fear by sharing your thoughts. Chances are, you will gain support you didn't know you had.

Finally, remember this: *You have nothing to lose.* They are not going to take your kids from you (even if you sometimes want them to), fire you, demote you, or cut off your little finger. You are attempting to gain a promotion. If you don't get the promotion, you didn't *lose* a thing. If fact, you still gained valuable experience, wisdom, training, and knowledge. Depending on how you handle it, you may also gain respect. Measure success by more than just a promotion. *You cannot always control the outcome, but you can always control your attitude.*

Bestselling author Charles Swindoll wrote this about attitude:

> *The longer I live, the more I realize the impact of attitude on my life. Attitude to me is more important than facts. It is more important than the past, than education, than money, than circumstances, than failures, than successes, than what other people think or say or do. It is more important than appearance, giftedness, or skill. It will make or break a company... a church... a home. The remarkable thing is, we have a choice everyday regarding the attitude we will embrace for that day. We cannot change our past... we cannot change the fact that people will act in a certain way. We cannot change the inevitable. The only thing we can do is play on the string we have, and that is our attitude... I am convinced that life is 10% what happens to me and 90% how I react to it. And so it is with you... we are in charge of our attitudes.*

Now let's talk about nerves. Nerves are obviously a symptom of fear so we can apply everything we used to fight fear to also eliminate nerves. Another way to get rid of nerves is to approach the assessment center as a day at the firehouse. Do not look at the assessment center as a test but just another day at work. Look at the building as a firehouse. Look at the assessors as peers, crew members, or other chiefs from a neighboring department with whom you are training.

Most of the exercises will place you in the role of the lieutenant, captain, or chief officer position for which you are aspiring. By mentally being in that position and looking at the people involved as members of the department or public, you will not be *testing* but simply performing your job. The result will be a much more calm and normal approach. Most of us are not nervous at work in the firehouse (hopefully), so if we mentally place ourselves in that place and role, we will reduce our nerves accordingly.

If you were being evaluated for a promotion, which would rather do: take a test or have your best day at the firehouse? You would probably choose the firehouse. We're much more comfortable in the firehouse and less nervous. We don't feel we are being *evaluated* in the firehouse. Take that mentality to the assessment center.

As we will discuss later in detail, the exercises will place you in the firehouse or with the public or on the scene of an emergency or at a drill, etc. The sooner you place yourself there as well, the less nervous and more comfortable you will be.

Another way to eliminate nerves is to simply have fun. Remember, you have nothing to lose and *everything* to gain. Go in with a carefree attitude (read the Swindoll quote on attitude again). Enjoy the opportunity to perform.

One story that illustrates this attitude is from the Super Bowl. In the final minutes of Super Bowl XXIII between the San Francisco 49ers and the Cincinnati Bengals, the 49ers were attempting to march down the field for a final game-winning drive. They were behind, and needless to say, the offense was a bit nervous as the entire world was watching.

While inside the huddle, quarterback Joe Montana looked up at the crowd and noticed actor John Candy in the stands. He laughed and said to one of his teammates, "Hey look . . . There's John Candy." The rest of the offense directed their dinner-plate-sized eyes at the stands to see John Candy. Once they realized that their quarterback was having fun by noticing a comedian in the crowd, they knew they would win. Their nerves went away, and they marched to victory.

Firefighting is the greatest job in the world no matter what rank you are. You should enjoy every minute of it, whether in a promotional process, venting a roof, or having dinner with the crew. Many people would die for this job, and many have. Remember that and be grateful you are a firefighter. Enjoy everything about it.

Check Your Baggage at the Door

Sometimes our attitude and nerves can come from unnecessary baggage. By the time most of us reach the season of our career where we are attempting promotion, we have accumulated some baggage. Baggage is anything that weighs us down and prevents us from having a healthy attitude going into the assessment center. Baggage can be work-related or home-related. We must free ourselves from it and check our baggage at the door before we go into the assessment center (see fig. 1–3).

One example is a candidate who was angry at the fire chief. During his interview, he told the panel that the chief was a liar. Needless to say, he didn't pass. This candidate would have made a great captain, but his baggage prevented him from reaching that goal. Later, he was promoted and is a very respected captain today. Part of his success was that he learned to check his bags at the door. Examples of baggage include the following.

Work

- Numerous assessment centers without promotion
- Anger with the fire department or member of staff
- Anger at the union
- Upset about recent contract negotiations
- Worried that the younger kids are passing you by
- Worried that a specific individual is going to score better than you
- Wanting to promote so you don't have to work for someone
- Looking incompetent
- Failure
- Feeling that the testing process is flawed
- Feeling like the testing agency is incompetent
- Bad experience from the last test
- Not liking an assessor
- Feeling that the department already picked who they want

Personal

- Feeling that you will let down your spouse and children
- Divorce
- Financial problems
- Missing a vacation to take the test
- Just fought with your spouse
- Kids are having problems in school

The possibilities are not limited to these lists. You are the only one who can determine—and more importantly—*eliminate* your baggage. Take some time to think about it. Chances are, there are things you may not have thought of. Do not let them surface in the assessment center. Check them at the door. Once again, it's about the one thing you can control in life: *your attitude.*

Fig. 1–3 This poor candidate has some baggage. He needs to check it at the door before he goes in. The sooner the better when it comes time to prepare.

Assessment Center Orientation 2

An assessment center simulates a day in the life of the position being tested by utilizing a series of exercises. The exercises require you to exhibit your ability to do the job. To determine the exercises used, testing agencies break down the job tested into behavioral dimensions. The dimensions categorize the KSAs of the job into major groups.

For example, *fire captain* could be broken down into three main categories: leadership, management, and emergency operations. Each of these main dimensions would have subsidiary KSAs associated with them. *The KSAs are the foundation on which exercises are built and scoring criteria are evaluated. So, you can see that developing excellent KSAs will prepare you for excellent performance in the assessment center. The problem is that most candidates do not make the connection between KSA development and test performance. They simply study or practice tricks rather than establish a plan to practice and develop sound KSAs.*

How Does a Testing Agency Know What the KSA and Behavioral Dimensions Are?

Most testing agencies utilize a job analysis questionnaire of some type. They give each of the current officers a questionnaire that asks what they do, how often, and how important a particular task is. Another question often asked is whether the task or skill is expected before or after promotion.

For example, the use of a staffing program by BCs is often learned on the job (after promotion), so it may not be utilized as part of the test. That being said, you should know how to use your department's staffing software if you want to be a BC in a department that requires BCs to perform staffing. *The goal is not to guess what is on the test. The goal is to be prepared for the job.*

In addition, job descriptions are reviewed and interviews are conducted with department officials to determine the most appropriate and relevant exercises. The scenarios should create a realistic environment for the position *and* agency test. Each fire department has different responsibilities associated with each rank.

An assessment center should not have exercises that are unrealistic for the position or the agency. For example, Las Vegas would probably not have an emergency simulation that involves a shipyard fire. In like fashion, a BC assessment should not include how to deal with a tough engineer in your firehouse, since that would be a first-line supervisor's issue (captain or lieutenant). If the captain or lieutenant then had an issue or couldn't handle it at his/her level, then the BC may intervene more directly. Similarly, a captain exam should not have a scenario in which you must pump a fire engine. However, some fire departments *do* require the company officer to pump an engine. So the exercises are truly aimed at creating a realistic challenge that is unique to the particular agency and position. *The key is to know the position you seek in your organization!*

What Kinds of Exercises are Used?

Although there are many variations, the most common fire service assessment center exercises include the following:

- The emergency scene simulation
- The oral presentation
- The role-play/counseling exercise
- The structured interview
- The in-basket
- The modified in-basket
- The written essay/report
- The supervisory exercise
- The leaderless group

We will discuss specific techniques and tools for each of these exercises in much greater detail later in the book. Many exercises are related, thus enhancing the realism of your experience by creating a true "day in the life of" the position being tested. A key point here is to identify related issues and utilize the common threads to act appropriately in your exercises. The more you respond as an actual officer, the greater your score.

For example, you may have an in-basket exercise with an e-mail from the training division stating that one of your captains is not completing his reports in a timely fashion. Later that day, you may have a counseling exercise with your *same* captain in which you must discuss his bad attitude toward the department. It might make sense to identify the earlier training records issue as a symptom of a greater problem that you must now solve.

Other common exercise connections would be the writing of a report for an emergency simulation exercise or an oral presentation based on a written essay. Once again, look for connections that help you enhance the realism of your performance.

Who are the Assessors?

Assessors are often individuals one to two ranks above the position being tested. They often work for another agency to maintain objectivity and some anonymity. Other individuals may include public administrators, civic leaders, or those with relevant expertise in the behavioral dimensions. Assessors typically go through a day of training to get oriented to the test exercises, your agency policies, procedures, and general expectations of the fire chief.

The most important thing to remember is that *if you are prepared, it does not matter who is assessing you.* You may cater a particular response to take into consideration the array of people who are assessing you, but your KSAs are not dependant upon who is assessing them. By most any yardstick, you will either have the skills or not. Do not waste time worrying about things you cannot control. If you have confidence in your abilities, it does not matter.

As shown in fig. 2–1, another person you are likely to encounter is a proctor. Proctors do not score you. They typically work for your fire department or for the testing agency. Proctors facilitate the process by reading directions, rotating candidates on time, supporting assessors, etc. Although they do not score you, you should remain in a professional testing mode around proctors, assessors, and anyone you encounter, whether en route to, during, or returning from the assessment center.

Fig. 2–1 Typically, assessors will be dressed in Class A uniforms, as shown on the left. On the right is a proctor, typically dressed more casually.

How do I Prepare?

Quite simply, learn the job. Nearly every book, expert, and testing agency will tell you that the best prediction of success in any assessment center—regardless of profession—is the candidate's *long-term preparation for the job.* In other words, you cannot fake it or take a shortcut. You must diligently prepare for the job you seek.

The KSAs to be an excellent officer require planning, hard work, and time to develop. *As you develop these KSAs, you will be preparing for the assessment center by default.*

Once promoted, many new officers are forced to learn on the job, by trial and error. Unfortunately, the errors can have catastrophic results.

It is incumbent upon you to evaluate your current situation, establish a plan, implement it, monitor the results, and adjust as needed.

Finally, the best way to approach an assessment center is to act like it's just a day at the office and you are already promoted. We will talk more about that a bit later.

What an Assessment Center is Not

An assessment center is not just an interview or a written test, although those may be a part of the overall process. Assessment centers force you to exhibit your ability—or lack thereof—to do a job. *You must perform.* In an interview, it is very easy to *say* that you have good decision-making skills, are good at organizing things, and are good with people. In an assessment center, you cannot fake it if you don't have it. It's that simple.

Unless you are an academy-award-winning actor, you will not fool anyone. Remember, actors have special effects, stunt doubles, editing, and the ability for a second take. This is not the case in the job or the assessment center. No one is going to come in and write for you, speak for you, solve problems for you, or smile for you.

The stress, realism, and very nature of assessment center exercises prevent you from winging it or covering obvious weaknesses. You will simply show your stuff, whether you like it or not.

With the right plan, an assessment center is also not impossible to conquer. If you have weaknesses, you can improve them; but know that it will take some time. Many candidates have failed numerous assessment centers, yet some ultimately succeed and get promoted because they have improved upon their weaknesses. *The key is accurately identifying your weaknesses, and improving them through developing and implementing an effective plan.*

The Mentality 3

If there is one secret to an assessment center, here it is: *Be the position.*

Make a mental mind shift from being test-oriented to being position-oriented. In other words, if you are taking an assessment center for captain, introduce yourself as "Captain Jones." A BC candidate should introduce himself as "Battalion Chief Smith." This has had very positive results in many assessment centers. When done in a respectful and enthusiastic way, it sets the tone and helps establish your mindset (see fig. 3–1). Remember, the exercises place you in the role, so the sooner you place yourself into that mindset, the better you will perform. The assessors will respond positively and see you in a fresh light.

Fig. 3–1 Notice the confidence in this candidate as he introduces himself as the lieutenant. Remember, only practice this technique if you truly feel that you have made the mental paradigm shift.

Many students ask, "Do I actually say *Captain* Smith?" The answer is yes, but here's the catch. If you are not comfortable introducing yourself as Captain or Battalion Chief Jones, then do not. Beyond these simple gestures, *you must truly believe, that for your assessment day, you have already been promoted to the position for which you are testing.* Introducing yourself as though you have the position is very effective, but only if you believe in yourself and can perform to back it up. Consider your comfort level in introducing yourself as the litmus test to your mental state. If you are comfortable introducing yourself as the officer, then you have truly made the mental shift for the day.

An assessor recently spoke of a candidate for BC. He said the candidate, "Came in as a captain and left as a captain." Consequently, he was not promoted. The assessors must see that you are ready for the job you seek. You must hit the ground running or you may be left behind. By making the mental shift, you instill confidence in the assessors, and they will feel comfortable giving you increased responsibility.

Do not enter the assessment center like you are taking a test. You will be more nervous, worried about what the assessors think, and be less like a leader. Your mindset must be in the role, without anything to lose.

So many candidates want to know what is on the test, what exercises will be used, who the assessors are, and what criteria will be used. In other words, they obsess about the test without just having confidence in their abilities.

Some assessment centers even spell out every one of these factors ahead of time, and candidates *still* fail. Many BC tests even allow the candidates to take home portions of the exam, and they fail anyway. So, knowing the questions ahead of time does not always help. *Your success has nothing to do with the test; it has everything to do with your abilities.*

Remember our Combat Challenge analogy? What good would it do for an out-of-shape person to know exactly what the Combat Challenge events were if that person did not have the skills and abilities due to a lack of exercise before the challenge? None.

Consider this mental preparation as your locker room pre-game. Your mindset must be in place at least 24 hours before your test. This means getting your spouse, roommate, or friend to join in the preparation. When you wake up in the morning, get them in on the role-play. Ask them to ask you how your new promotion has been going. Tell them that it's great and that you are looking forward to another great day at work. This warm-up time will get you ready for your assessment. By the time you enter the testing environment, you will be far into your role. You will likely be more comfortable in role-plays, oral presentations, and simulations since you will be warmed up mentally.

Excellent officers do not go into work worrying what calls they will respond to, if they will have to write a report, or afraid that they will have to use people skills. Excellent officers go to work ready for anything, full of confidence. They have spent years developing their craft and do not think twice about taking care of business. In like fashion, you must go into the assessment center brimming with confidence (but not cockiness), ready to take on whatever comes. That's what a great officer does.

If you do not have this confidence, then you must honestly assess your weaknesses and improve them through planning, effort, and time. We will address setting up your plan later in the book.

Mental Readiness/Locker Room Preparation

The final day(s) leading up to the assessment center are just as important as the months or years you have spent developing your KSAs.

Like a football team that is ready to take the field, you must mentally prepare for your performance. A good football team has developed and honed the basics, practiced over the long term, and conditioned themselves to perform a certain way. You have also developed, practiced, and conditioned your KSAs.

As you approach your assessment center day(s), you must mentally prepare so that you get the most out of your KSAs. Think of this as your locker room pre-game preparation. Following are three ways to mentally prepare.

First, go to the testing site before the test takes place

Do not go to the testing site during the process when others are testing. The best time to go is a few days before the entire process has begun.

A word of caution: You may be specifically instructed not to go to the assessment site at any time before your test day. If instructed not to go, then follow those instructions.

Walk around the site. Mentally picture yourself arriving, looking sharp, feeling confident, and ready to go. Find where you will park and familiarize yourself with the route from the parking lot to the actual testing site (see fig. 3–2). Do not attempt to enter the building; just get a feel for the site, parking area, and general location.

Fig. 3–2 Walking around the site several days before the process starts will help you get acquainted with your surroundings before game day. Picture yourself arriving confident and ready to take on whatever comes!

As you familiarize yourself with the site, you can picture yourself as the captain, BC, etc. Since you will picture this as a firehouse, you can become acquainted with your surroundings and make them more familiar. When the day of your test comes, you will have a greater sense of calm and familiarity.

As you walk around, tell yourself, "I will shine here . . . This is going to be my time . . . I can't wait." Choose whatever positive phrases work for you. Like a football team, you can picture yourself on the field and performing your best.

Second, become the position

As we discussed before, your spouse or significant other can help you get mentally promoted. You want to go into the assessment center as though you already have the job, are the officer, and are simply going to another workday. Truly feel as though you have nothing to prove and nothing to lose—just like a real officer.

This could involve you talking to your spouse about the job in terms of having actually been promoted. Here are some examples:

- "I love my new job as battalion chief."
- "This new position has been great."
- "Thank you for all of your help in me getting promoted to captain."
- "They want me to talk to some chiefs from another department today about how we fight fire."
- "I'm supposed to give a presentation today to some chiefs."
- "I think today is going to be a busy day in my new job."
- "I'm attending a training session today."

By speaking in terms of another day on the job in your new position, you mentally promote yourself from test candidate to officer. This will boost your confidence, make you more assertive, and give you command presence.

Third, find your special inspiration

Reach deep down inside and look for something or someone who inspires you. What or whom can you think of that will give you that extra sparkle in your eye? Perhaps your spouse, your kids, or other family members will give you that inspiration, as shown in figure 3–3.

Perhaps faith or thinking of a dad who is/was a firefighter will give you that feeling. Some may find inspiration by thinking of a friend on the job who died in the line of duty. Another choice may be an instructor, a mentor, or an officer who influenced you. The choice is very personal and up to you.

Try carrying a picture of the person who inspires you and listening to some of your favorite music on the way to the assessment center.

Fig. 3–3 Our family can be our greatest source of inspiration.

Offensive vs. Defensive Performance

Most candidates go into the assessment center with a very defensive posture. Candidates hope to give assessors what "they are looking for" rather than giving the panel their best. These candidates attempt to merely survive the test, rather than conquer it. This defensive posture can create nervous anxiety, cause indecision, and prevent you from being the confident and capable officer that you could be.

This defensive posture is also very easy for assessors to see, as shown in figure 3–4. Candidates squirm in their seats, make uncertain comments and decisions and are easily rattled with questions by the assessors. None of these qualities inspire confidence in the candidate. They only trigger doubt in the assessors.

Fig. 3–4 Would you follow this meek candidate into battle?

Much of the assessment center scoring criteria is subjective. The assessors' opinion on your behavior is just that, an opinion. Subtle nuances of confidence, command presence, assertiveness, and discernment are often difficult to assess. The assessors yearn to be inspired. They do not want someone who gives them a canned answer as to what they *think* is right. Assessors want to be inspired and confident in the people they are scoring. They want someone who is ready for the job, capable in their abilities, and confident in their decisions, as shown in figure 3–5.

Fig. 3–5 This inspiring candidate is ready for his new job.

Assessors are charged with the responsibility of choosing the future leaders and officers of a fire department. These decisions have life and death consequences that assessors do not take lightly. A meek and insecure candidate will not miraculously make a great officer.

Resist making silly comments to the assessors, whether before, during, or after an exercise. One candidate saw an assessor in the bathroom during an assessment center and was asked, "How are you doing?"

The candidate answered, "Lousy, this stinks!"

Don't make nervous comments about being the first candidate after lunch, the last to go, the first to go, etc. The assessors don't care about when you are going, so it doesn't matter. Don't worry about if they are tired, bored, hungry, etc. You cannot control it, so don't devote any energy to it. It's unprofessional and will detract from your performance.

You cannot fake confidence. As stated before, confidence is the cart that follows the horses of time, education, experience, and trial and error. Learn from your experiences and improve on them.

Once again, it's all about the KSAs.

Beyond confidence lies an offensive and inspirational posture of performance. By *offensive* we do not mean to offend the assessors or attack them. We simply mean that you take on the assessment center with confidence, vigor, and a bring-it-on attitude.

Inspiration can come from simply having passion for the job of firefighting. The assessors are going to see many candidates that fit a mold and have a defensive posture. Go beyond surviving the assessment center; master it with an inspiring attitude that polishes your KSAs and makes you shine.

While *never cocky and always respectful*, inspirational candidates are truly officers. An inspirational candidate leaves the assessors wishing he worked for their department. Inspirational candidates teach the assessors some new things in the process. Perhaps your in-depth knowledge of the labor-management process introduces them to relationship by objectives (RBO). Maybe your passion for team building shows them a new way to inspire a crew. Possibly a way you treated a member of the public who was irate gave them a tingle down their spine. Whatever the case, take on the assessment center as a confident, capable, and inspiring officer!

The KSAs To Be an Excellent Officer 4

As we discussed earlier, the KSAs to be an excellent officer are the same KSAs evaluated on the test. When reviewing these elements to being a successful officer, honestly consider your own level of skill.

If you are lousy at oral communication, that's okay. What matters is what you are going to do about it. You can improve, but it will take time.

Remember, just knowing these definitions does not make you good at them. Your knowledge of the definitions will give you a goal from which to aspire and a direction from which to start.

We are all given a certain package of innate gifts. Some of these KSAs are naturally part of who we are. Others will take more work to develop. *The key is to bridge the gap between where you are and where you want to be.* Do not underestimate the opportunities in your personal life that can enhance your KSAs. Owning a business, having kids, coaching a baseball team, and keeping a budget are all opportunities to enhance some skills.

The KSAs for this book have been broken down into three main dimensions: leadership, management, and operations. Leadership deals with the *people* element of being a fire officer. We lead people but manage things. Management consequently deals with the *stuff* element. Finally, the operations dimension deals with the *calls*, where people and stuff come together.

When looking at each of the KSAs, we will break them down further into three areas: relevance, thoughts, and questions.

Relevance will help us to understand why the KSA is here in the first place, what makes it important, and how fire officers use it.

Thoughts help us to understand what the KSA should look like when performed properly.

Questions will help us assess where we are and formulate our plan for bridging the gap. Remember, varying forms of these KSAs are all evaluated in an assessment center.

Leadership

Leadership has been called getting the organization's work done through *people*. Remember, leadership is about people (see fig. 4–1). People are our most important asset. Leadership skills can be learned. Do not underestimate your ability to learn these KSAs.

Fig. 4–1 Leadership means listening, supporting, motivating, and many KSAs that take time to develop.

Interpersonal skills

This refers to the ability to empathize with others. It also includes patience, listening, understanding, and emotional intelligence.

Relevance. The fire service is 100% about people helping people, 100% of the time. That's why we list it first. This is the most important skill you can have and the most often relied upon. As an officer, you will do less with your hands and more with your words and actions.

People skills are tested every day internally and externally. Internally, you will use people skills with a member of your crew, another company, another shift, or the finance division. Externally, patients, family members, bystanders, homeowners, teachers, parents, law enforcement, and countless others will cross your path. Assessment centers always assess interpersonal skills.

Thoughts. This is a tough one. Personalities are very different. Interpersonal skills are more about respecting someone than liking someone. Officers can be fun to be around but disrespected for their lack of ability. Conversely, extremely knowledgeable officers can be disrespected for their arrogance. There is a balance.

Strong, positive, and highly respected personality traits for fire officers include the following:

- fair
- patient
- sensitive
- empathetic
- calm
- consistent
- positive
- helpful
- understanding
- solution-oriented

Questions:

- Do I have a hard time being patient with people?
- Do I expect everyone to conform to my way?
- Am I often misunderstood?
- Do people respect me?
- Do I earn respect or demand it?
- Do I admit when I am wrong and learn from my mistakes?
- Am I easy to get along with or hard to be around for 24 hours?
- Am I firm, fair, and friendly?
- What mentors can be an example to improve my interpersonal skills?

Oral communication skills

This refers to the ability to *seek first* to understand and then to be understood.

Relevance. As an officer, you will spend more time communicating orally than anything else. Everything from training, phone calls, public education, crew interaction, emergency scene orders, and initial size-up all rely on oral communication. Many people who lack other skills make up for it with oral communication (the gift of gab). Also, the single most rampant problem at emergency scenes is lack of effective oral communication.

Thoughts. Good oral communication skills start with listening.

Listening. First, always seek to understand the person who is communicating with you. Good listeners send the message that they care. Listening is more important than speaking. We have two ears and one mouth for a reason. We should listen twice as much as we speak. Give feedback with gestures. Let the sender know that you either understand or do not understand the message.

Speaking. Communication is a skill and an art. Conveying a clear, concise and understandable message takes effort. Speak in terms that your listener will understand. Do not use technical jargon to impress people. This will only demonstrate your insecurity—not your knowledge. When speaking, slow down and ensure that your listener understands the message that you want to send.

Phoenix Fire Chief Alan Brunacini has a saying: "Be sincere, be brief, and be seated."

Questions:

- When I communicate, do I seek first to understand? Or do I wait for them to shut up so I can get my point across?
- Do I *really* listen? Or do I think of my next point while the other person is talking?
- Do I often feel misunderstood?
- Do I keep it simple when I speak? Or do I like to hear myself talk?
- Do I use radio traffic wisely?
- Would taking a speech class help my skills?
- Would conducting station tours, schools presentations, company drills, and inspections help to calm my nerves in front of people?
- What else can I do to improve my oral communication skills?

Written communication skills

This refers to writing in a clear, concise, and understandable format.

Relevance. Written communication is part of the job as an officer. We write reports, narratives, memos, and e-mails. We fill out many logbooks and forms. Reports can be subpoenaed, and you may have to refer to a report in court that you wrote years before. As a chief officer, you will find that budgets, committee reports, and plans become a part of everyday life.

Thoughts. Like oral communication, written communication seeks to be understood. Since you do not have the luxury of feedback, you must write as clearly as possible and anticipate any questions that the reader will have. Take into account the reader's perspective (or lack thereof). What is the goal of your document? Do you want to educate, entertain, solicit input, gain support, or gain approval for a project?

When writing, follow some simple rules.

- Tell them what you want to tell them (intro), tell them (body), and tell them what you told them (summary).
- Write in terms the reader will understand; educate if necessary.
- Organize your thoughts with an outline before you write.
- Utilize a thesis statement at the end of the introduction. (For example: "Therefore, the fire department should purchase tractor-drawn tiller trucks from this point forward.")
- Use topic sentences at the beginning of each subsequent paragraph to support your thesis statement. ("The first reason tiller trucks would be a benefit is....")
- Conclude or summarize by restating the main points. ("In summary, tiller trucks would benefit Dream Fire Department in three ways.")
- Write within your ability. Do not use words that you cannot spell or cannot define.
- Adhere to rules of spelling and good grammar.

Questions:

- Am I confident in my writing skills?
- When was the last time I wrote something more than a page long?
- Was this the first time I have heard most of these simple rules on written communication?
- Is there a book I can buy that will help me with some rules?
- Is there a class I can take?
- Who do I know that is good at writing?
- What else can I do to improve my writing skills?

Motivation skills

This refers to finding individual needs and desires in order to boost internal motivation.

Relevance. Motivation is one of the most important skills to the fire officer, regardless of rank. Motivation is a derivative of interpersonal skills, but it is so important that it warrants its own section.

As fire officers, people become our tools. Unlike a saw that needs fuel, we need to *find* the fuel that our people need to start and run. As firefighters, we almost solely rely on task-level skills to do our job. As company officers, we still rely on task-level skills like cutting holes and pulling hose; however, we start to rely on our people to be our tools as well.

Finally, at the BC officer level and above, we solely rely on people as our tools. As we all know, a paycheck is why most career firefighters show up to work at 8 A.M.; what they do when they get there is based on *their* internal motivation.

Thoughts. Every individual has a unique set of motives. Do not be fooled by our uniforms. We may look the same on the outside, but everyone is amazingly different on the inside. The individual determines their motivation. Your job is to find out what that motivation is and then utilize it to get the individual enthused to do the job.

Remember that, unlike a chainsaw, which always takes fuel, people have emotions, moods, good and bad experiences, and a host of other factors that make them unique. You cannot simply read the owner's manual on each person to find out what fuels him/her. You must find out each individual's fuel through time, caring, talking, observing, and team building.

One of the worst things we can do as officers is to inflict our motives on our troops. That will turn them off as well as create resentment on their part and frustration on your part. Don't ask yourself, "Why won't he do what I want?" Instead, ask, "What will make *him want* to do it?"

Questions:

- How much experience do I have working *through* other people (different than working *with*)?
- Have I ever had to motivate someone? How did it go?
- Do I expect everyone to be motivated by my values?
- Do I seek out what motivates others?
- Have I ever even thought of this before?
- What projects can I lead in order to help people get motivated in completing them?
- Are there classes I can take?
- What else can I do to improve my motivational skills?

Ability to initiate

This refers to recognizing what needs to be done and then stepping up to do it.

Relevance. Leadership means initiating action when no one else is. Many people with bugles are afraid to step up and lead. If you simply wait to be told what to do all the time, you are not a leader. Remember, leaders initiate, set course for uncharted territory and inspire.

Initiative is important to the fire officer in countless areas. You may have to make a command decision that is unpopular by having a wet drill at night. Perhaps you must take command because the BC is out of the area or organize companies on a division of a fire or give direction for a drill. Many officers wait for someone else to do these things because the officers lack initiative.

Thoughts. Initiative takes effort. You may not like the spotlight, not want to be in charge, not like attention, or not like to give direction. If that's the case, don't promote. Why? Because to be a good officer, you must have initiative.

This is one of the most rampant problems with today's fire officers: they lack initiative. Look at the lack of initiative in others as an opportunity for you to influence the direction of an incident, drill, project, or decision. *Develop this through getting out of your comfort zone.*

The troops in the fire service long for strong leadership. Give it to them. *A good rule of thumb is to go until someone tells you to stop.* Check with the boss to see if you ticked anyone off. Check with the troops to see if they like what you are doing. It they both give you the thumbs up for your initiative, keep going. Not everything is listed in a memo, bulletin, policy, or procedure. Sometimes you must just do what you think is right when no one else does. That's leadership through initiative!

Questions:

- Do I step up or step away?
- Have I ever stuck by an unpopular decision because I knew it was the right thing to do?
- If I was a company officer and my engine drove by a person with a flat tire, would I tell the engineer to pull over?
- What needs to be done at my workplace that no one is doing?
- Will I do it?

Team-building skills

This refers to creating a unified team comprised of individuals who aspire to accomplish a common goal.

Relevance. Firefighting is a team sport. We must always remember that the individuals are only as successful as the team. If the captain and firefighter make a great hose stretch to the second floor garden apartment and wait for water too long due to the engineer's incompetence, then the team lost the battle.

Officers are in the team-building business. We build teams in the firehouse, on the fireground, at drills, in battalions, on shifts, and even on our days off. In our business, we cannot afford to have a dysfunctional team. Everyone must be on the same page.

Thoughts. Like a coach who puts the right people in the right positions, fire officers must position their folks according to their strengths and weaknesses. Position your people for success. This has little to do with rank or formal position.

For example, if your senior firefighter is a bit low on self-esteem since he flunked the engineer exam, have him show the rookie (probie) how to start the chainsaw. He wins because he gets to show his stuff, the probie wins because he gets mentored, and you win because you are free to take care of other business. The team has been positioned to win. You must build trust, cohesion, communication, and morale.

Teams have a personality that is a conglomeration of the individual personalities on the team. Every fire department has a "B shift" (no offense to you B shifters) and every battalion has the *black sheep* company or a *cowboy* for a captain. Wouldn't you know it, but the rest of the team will follow that captain's lead. No one has more influence on the personality of a team than the leader. Whether a company officer or BC, you have tremendous influence on your team.

As an officer, you must determine the type of team you want based on what you have (people, resources, and challenges) and where you want to go (the goal). The key is to get your troops in on the vision. Empower them and give them ownership. Find out their motivation and fit them into the team accordingly. *If you fail to build the team in the fashion you desire by utilizing your troops' strengths, your troops will build their own team and you may be "voted off the island."*

Questions:

- Have I ever helped build a team, or have I just been a member of teams?
- What was good or bad about the teams that I have been a part of?
- How would I have improved my team?
- Have I ever been voted off the island?
- What are my current team's strengths and weaknesses?
- What can I do with my crew tomorrow?

Delegation skills

This refers to working through your people to get things done.

Relevance. Don't be a ball hog. Delegation is required by fire officers to get all the work done each day. The benefits are many.

- First, you develop your people by allowing them to learn new skills.
- Second, you empower your people by giving them a say on how things are done.
- Finally, you free your time to focus on other issues that need *your* attention.

Thoughts. Delegation, like initiative or motivation, takes effort and skill. You must often take the time to slow down and think of who would be a good choice for a particular task. Most officers would rather "Do it myself" than take the time to delegate.

Refusal or inability to delegate has three negative results.

- First, you become overwhelmed since you try to do everything yourself.
- Second, your troops get ticked off because their one-man-show of a boss doesn't think they can handle anything.
- Third, no one wants to help since you try to do everything anyway.

It is pretty easy to see officers who are lousy at delegation. They run around like a chicken with its head cut off while their troops point and laugh.

Think of an orchestra conductor. She sits in front of the orchestra, baton in hand, effortlessly leading the team. She doesn't run around trying to play the violins, hit the drums, or clash the cymbals. Those tasks are delegated. Another word for someone who won't delegate is *micromanager*. We all know how much *that* word is "loved" in the fire service.

According to Chief Brunacini, firefighters want five things:

- Tell me what you want me to do.
- Train me how to do it.
- Give me the tools.
- Get outa my way.
- Tell me how I did.

This is delegation.

Questions:

- Am I a micromanager?
- Do I enjoy delegating?
- Do I over-delegate to get out of work?
- Am I threatened by others' opportunities?
- Do I look for opportunities to develop my folks through delegation?
- How can I go from a doer to a leader?

Empowerment skills

This refers to building your people and the team through sharing ownership.

Relevance. Empowerment goes hand-in-hand with delegation. You really must have both to be most effective. Firefighters hate being told they need to do something yet not given any latitude in how it's done. Empowerment builds trust. You demonstrate your trust that your troops have good judgment and skills, and they demonstrate their trust by exceeding your expectations.

Thoughts. Like delegation, you must choose the right person for the job and give the appropriate amount of empowerment. You must be willing to live with the decision the troops made, especially if it is different from what you would have done. This demonstrates your faith and flexibility. Empowerment leads to buy-in and ownership for the project at hand. Opportunities to empower and delegate are around every corner. Take all the opportunities you can to gain that trust, ownership, and buy-in, then build your team and your future leaders. A good rule is to "train others to replace you." Someone did it for you, so pass it on by mentoring and empowering.

Questions:

- Do I take my ball and go home when I don't get my way?
- Do I have to call all of the shots?
- Can I trust others easily?

- Do I fake empowerment only to steal it back, late in the game?
- How can I trust others better?
- Who did this for me, how did it help me, and how can I pass that on to others?
- Who am I training to replace me?

9 Ability to remain consistent

This refers to providing stability through continuity of actions and philosophy.

Relevance. No one likes having to guess what today's rules are. As an officer, your troops want you to be consistent in your decision-making style. Although each situation is unique, we must always attempt to apply a relatively standard problem-solving style to our actions. "You never know what he's going to do" is not a flattering comment for a fire officer. In the world of ever-changing environments faced by today's firefighters, any continuity can provide necessary confidence and stability, thereby minimizing the effects of changing conditions.

Thoughts. "Do as I say, not as I do" is a great way to ruin your credibility. We must remain consistent in our thoughts, philosophies, and actions. You cannot chew out a crew for not wearing the proper uniforms and then be caught wearing your sandals and shorts at the dinner table. Such trivial topics will undermine trust and affect far more important issues like whether to go to the roof or not.

That double standard may transmit into, "He'll send us to the roof, but I bet he wouldn't get up here. We're not going!" Such thoughts can create doubt in your commitment to the welfare of your troops. Would you send them into harm's way from the comfort of your BC rig? They may think so if you told them to wear their uniforms while you wore your shorts.

Questions:

- Do I practice what I preach?
- Do I vacillate with the popular opinion? Or do I remain consistent?
- Do I employ a relatively standard problem-solving sequence? Or do I fly by the seat of my pants?
- Where did the expression, "fly by the seat of my pants," come from anyway?
- Have I worked for someone who was inconsistent? If so, how did that make me feel?
- How can I be more consistent?

10 Ability to put others first

This refers to developing consideration for others' needs, feelings, and motives before your own. It includes having empathy.

Relevance. The ability to put others' needs ahead of yours and put yourself in their shoes is another derivative of interpersonal skills. Once again, this warrants further discussion. Very few firefighters enjoy working with officers who are self-centered and continue to put themselves ahead of their crew. This attitude is easily seen in the captain who makes sure he's fed first and then doesn't do the dishes or the BC who expects her captains to do her reports for her. These traits will, in the minds of the troops, transfer into poor fireground behavior. How? Anyone who cares more about themselves than their troops is apt to put their troops in harm's way.

Thoughts. *One of the biggest secrets to excellent leadership is to get* behind *your people.* Sounds backwards, but it works. *People will only follow those they trust and only sacrifice for those who have sacrificed for them.* As leaders, we must be the first to sacrifice, without resentment.

A leader without followers is just taking a walk. To get the kind of "fall on a sword" followers that only few people have, we must put our people first. When you accepted your firefighter badge, you swore to protect the lives and property of the customers in your jurisdiction. When you accept an officer badge, you accept the unspoken need to put your troops ahead of yourself. That's how you protect them.

As we promote, it becomes more and more difficult to remember the teamwork, trust, and camaraderie that makes the fire service so great. Why? Administrative officers often lose touch with life in the firehouse: *A simple place where sobering daily reminders keep us grounded as to why we exist.*

It takes effort to put others first and to stay in touch with those we are charged with caring for and leading. Some captains have a hard time coming out of their office or dorm. They never spend time with the troops at the dinner table. Some BCs never visit their crews, just to say hello, and make sure everyone is okay. Most assistant chiefs never come out to the stations to "stick their toe in the water" or have a cup of coffee.

All of these things take effort and demonstrate that you truly care. Such effort, like delegating or empowering, will pay huge dividends of support and respect, and it allows you to see potential problems while they are still small and can be mitigated with ease. Think of the gardener who walks in the garden each day. The slightest weed can be easily detected and pulled. Waiting until the weeds are three feet high is too late, and it will be 10 times the work to pull out of the ground.

Questions:

- Do I put others first? Or do I make sure my own needs are met first?
- How do I feel when my boss clearly puts himself first?
- Will I repeat that cycle?
- Is there something I am missing that one of my teammates needs?
- What can I do to put my teams ahead of me, even if I am not an officer yet?

Now that you have been given the definition of the 10 KSAs of the leadership dimension, you have a foundation to build upon. Remember, just knowing these definitions will not guarantee that you will perform them as described. You must train, practice, and develop these over time, with effort and dedication to change your behavior.

Management

Our second dimension is management. Unlike leadership that deals with people, management deals with *stuff.* We manage things and lead people. Things don't have emotions. They don't get mad, hold a grudge, or gossip. Nonetheless, we must be good at managing things as part of our job.

Management involves planning, organizing, coordinating, controlling, evaluating, and a host of other components that we will explore in the following text (see fig. 4–2). Assessment centers love to assess these KSAs. You will notice that some leadership KSAs overlap with management KSAs. Officers do not operate in a vacuum. For example, problem solving is a management skill that could be applied to a person with an attitude. You may need to lead them to a solution.

Fig. 4–2 Managing things like information, resources, and time are critical to the officer, especially on the fire scene.

Problem-solving skills

This refers to assessing symptoms, defining problems, and implementing solutions. Consider and choose what solutions best fit the need. Then monitor results and make adjustments as needed.

Relevance. Being an officer is about solving problems. Problems may involve people, equipment, apparatus, customers, time, emergency operations, or all of the above. They could involve an acute problem like a fire or a chronic problem like a disciplinary issue. One of the biggest KSAs evaluated in an assessment center is problem-solving skills.

Thoughts. *The key to problem solving is to know the difference between a problem and a symptom.* Think of an emergency medical technician (EMT) doing a patient assessment. She looks at the signs and symptoms to define the problem. Only then can she apply an appropriate solution (treatment). Focusing too much on one symptom and rushing to a solution can be fatal. (For example, administering nitroglycerin for chest pain without looking at the blood pressure.)

A common mistake in life and in an assessment center is to focus too closely on the symptoms and not on the problem that is causing the symptoms. Only after closely *looking at all the symptoms* can we define the problem. We may need to *look at some options* of what the problem is first. Once we *define the problem*, we then *look at potential solutions.* Once we choose a solution, we then *implement and monitor* our progress. See if the plan is working. Then *adjust* as needed.

Let's look at an example involving a personnel problem. You have an engineer with a bad attitude toward you. He is undermining you whenever you try to train the probie. These are *symptoms*. With some problem-solving, discussion, and observation, you find that the underlying *problem* is that he has taken five captain exams. You were hired on your first test and are 15 years younger. He resents you even though he hardly knows you. The reason he keeps messing with drills is that he wants to show the probie that he knows his stuff more than you. He wants validation. The probie doesn't know about the five tests he failed. What would you do now? Instead of taking his actions personally—which could shut down communications and make things worse—you may now offer to help him promote by coaching him, letting him run drills, and helping him find classes.

Here is a checklist for review.

- symptom (engineer disrespecting you, especially during drills)
- problem (resents you for promoting quickly since he failed five tests)
- options
  - resent him back
  - tell him to get over it
  - tell him he should have studied more and not to make it your problem
  - help him promote
- solution (help him promote)
- plan of action (coach him on duty, find some classes to help him, let him run drills for experience and to build confidence)
- monitor (evaluate his skills and attitude)
- adjust (he is much more appreciative toward you but still is reluctant to take classes, so you keep encouraging)

Questions:

- Do I rush to a conclusion? Or do I take some time to find out what is going on?
- Have I ever been falsely accused of something? If so, how did I feel?
- Do I have a lot of problems in an area (promotions, people, policies, politics, etc.) and keep trying to blame someone (something) else?
- Am I the common denominator to my chronic problems? Or are they all just a big incredible coincidence?
- Can I conduct role plays with my crew to practice solving personnel problems?

Policy knowledge

This refers to knowledge of key policies and how to find other policies.

Relevance. Whether we like it or not, policies are part of life in the fire service. They help us stay on the same page and increase our safety and effectiveness. You must know and adhere to your department policies and procedures as part of good risk management. Be sure to remain consistent with department philosophy.

Thoughts. Although you don't need to recite every policy verbatim, you must know the key policies in your department and where to find all of them in case you ever need to look up something.

Some key policies include the following:

- harassment
- safety
- worker's compensation
- apparatus driving
- communicable disease
- discipline
- grooming
- uniforms
- code of conduct
- maintenance
- training
- report writing

Policies provide the boundaries for behavior. As an officer, you now go from just disciplining yourself to disciplining others as well. This is a big challenge for most new officers. We hate peer pressure and like to be liked. As a new boss, we are in a new role.

If you are waiting to follow policy until you are promoted, you are setting yourself up for an extra bumpy ride. Firefighters hate hypocrisy. Actions speak louder than words, and the "do as I say, not as I do" style will ruin your credibility. The time to start following policy is *now*—not when you promote. Your troops will think, "Why should we do what he says? He never did it until he promoted."

Questions:

- What policies are *key* in my organization?
- Where do I find them?
- What decisions would require me to contact my boss for approval?

Goal-setting skills

This refers to defining realistic, measurable, and attainable goals. Part of that process includes establishing sequential objectives, including (a) implement, (b) monitor, (c) adjust, and (d) accomplish.

Relevance. We must set goals to measure our progress and dedicate our resources in the most effective and efficient way possible. Most organizations, especially fire departments, are goal oriented. Goals let us see the big picture before we start looking at the pixels. Most officers fail to set goals to help accomplish tasks. In fires, we call them *tactical objectives*. We can also use goals to evaluate operations, training, apparatus, and maintenance. Just taking some time to establish and communicate clear goals for one shift will pay huge dividends.

Thoughts. Like problem solving, goal setting has some rules if you want to do it right. Goals must be realistic, measurable, attainable, and established through sequential objectives. Like problem solving, you must implement, monitor, and adjust if you want to accomplish. Let's look at these more closely.

Realistic means that we should only set goals that are relevant and that we actually have some means of accomplishing. There's a saying. "Don't try to teach a pig to sing. It only frustrates you and annoys the pig." Sounds silly, but we often hit our head against the wall trying to accomplish the impossible. Some people just can't do what you wish they could, and some things just won't happen.

Second, *measurable* means we must be able to track our progress so that we remain motivated and know how much we have yet to accomplish. We can then make the necessary adjustments.

Third, *attainable* (unlike realistic), considers the ability to accomplish the goal with the resources given in the allotted time frame.

For example, it *is* realistic for a 12-year-old boy to want to fly jet fighters in the Air Force, but it *is not* attainable in one year. He can set the attainable goal to fly jet fighters in the Air Force by the time he is 26.

Fourth, we break down goals into *objectives*. Objectives are the smaller bite-sized components of a goal. Objectives must be set in a sequential order.

For example, our kid who wants to fly jets must do the following:

1) get good grades
2) graduate college
3) enter the Air Force
4) enter the pilot training program

He can't enter the pilot program before he gets a degree. The Air Force won't let him. In the firehouse, we may sit down with our crew and establish some objectives that accomplish multiple goals.

For example, the department wants you to pump test your engine. Your firefighter wants to practice for the engineer's exam, and your engineer wants to promote to captain. Since you are a genius, you ask the engineer to coach the firefighter through the pump test. All three goals are met through your delegation, empowerment, and goal setting. Beautiful!

This is also known as management by objectives (MBO). MBO is a technique in which the boss helps the employee tie his/her personal goals to the agency goals. In the example previously given, the captain accomplishes the agency goal (pump testing) by tying it to the goals of the crew members (promotion). The crew members' motivation to promote accomplished the goal of pump testing. This is a win-win situation.

On the fireground, three levels of goals exist: strategic, tactical, and task.

- *Strategic* level goals are established by the incident commander (IC) and look at the broad large-scale end result (offensive attack, defensive attack, etc.).
- *Tactical* level goals break down the large goal into objectives (interior attack, rescue, ventilation, etc.).
- *Task* level goals are the actions taken to accomplish the objectives (pull 1½-in. attack line, start chainsaw, perform right-handed search, etc.).

Questions:

- What are my current goals?
- What objectives will I accomplish toward each of my goals?
- How can I solve some of my challenges (problems) in promoting and becoming an excellent officer by establishing goals and objectives?
- Are my goals realistic, attainable, and measurable?
- How will I measure them?
- What are some of my department's goals for the next year?

Time management skills

This refers to remaining on task in the allotted time frame. It also means planning ahead several steps and moving forward as needed. Good time management helps us accomplish goal(s) in the time allowed.

Relevance. Time is a resource. Time is one of our most valuable assets. We deal with time every day in the firehouse, on the fireground, in budgets, with staffing, response times, etc. You will be under significant time constraints during the assessment center. You only have 24 hours in a shift and only so long to safely perform an interior attack.

As emergency responders, we must work within tight time frames. If we are not successful, bad things can happen. As an officer, you must quickly take action, while maintaining safety, effectiveness, and efficiency.

Thoughts. Time does not change. It's a constant. One hour is one hour. What we do in that hour is the variable. We must actually switch from managing our time to managing our *self* and our *resources* more efficiently within the constant of the time given.

Prioritization is the key to effective time usage. As officers, we balance the critical needs of a situation with our resources. We must not forsake safety, but we cannot afford to wait until conditions are perfect before we take action. Conditions will never be perfect; that's why there is a fire. The window of opportunity closes very rapidly to safely perform interior attack, ventilation, and rescue.

You must have excellent instincts to quickly recognize a situation, develop a plan, and take action in a safe manner. These skills take years to develop. The previously discussed skills of problem solving and goal setting come into play. On the fireground, the time is compressed. We make high-risk decisions with low information and low time. In the nonemergency environment, we can practice some of these time usage skills while we have more discretionary time. However, you cannot possibly measure your success if you have no goals that you can measure against. See how all of these factors overlap?

Questions:

- Do I waste a lot of time?
- Do I get easily distracted?
- Do I start new projects before finishing old ones?
- Do I look for excuses to put off projects?
- Am I frequently late?
- Do I procrastinate?
- What can I do if I answered "yes" to most of these questions?
- Would it help me to do oral presentations that are timed?
- Do I wear a watch?

Prioritization skills

This refers to establishing an order of actions to take based on level of importance, resources available and time allowed.

Relevance. We will break down effective time usage a bit more by looking at prioritization. Time taken to prioritize will pay huge dividends in effective use of resources. It may feel like an eternity to stop, wait, think, and establish the best way to use your resources. But if it means setting up a truck in the right place the first time, it's worth it.

Thoughts. As stated previously, we must see the big picture so that we don't waste resources. For example, if you are in a boat headed for Niagara Falls, will you really worry about bailing out the water that is leaking in or would you jump out and swim to shore? By seeing the big picture, you would realize that there's no point in bailing out the water if you are going to plummet to your death. The priority is to jump out and swim for it!

Remember this formula: Goals + conditions + resources + time = priorities.

Questions:

- Do I rush into things? Or do I look at the big picture first?
- What are some tactical priorities on a defensive vs. offensive fire scene?
- What are my priorities now?
- What projects can I work on that will force me to set priorities in order to accomplish everything in a given time frame?

Resource management skills

This refers to controlling, maintaining, and utilizing resources safely, effectively, and efficiently.

Relevance. As officers, we have a myriad of resources available to us: personnel, time, apparatus, equipment, special services, policy, SOPs, facilities, training, employee assistance programs (EAPs), money, and our own KSAs to name a few. We must manage these resources with the three priorities we keep mentioning: safe, effective, and efficient (SEE). This term is often used as a verb. We must "SEE" through our problem. In an assessment center, as in life, you must know the resources available to you, when to use them, and how to use them in the best way possible. The best way is SEE.

Thoughts. Let's break down these three SEE priorities into areas of operational resource management. These can also be applied to training and just about any other facet of our business.

First, any time we take action or utilize a resource, we must be *safe*. Obviously, safety is always our first priority. We cannot function if safety is compromised.

Second, we want to be *effective*. Is our use of this resource effective? Is it making a difference? Are things better or worse since we employed this operation or resource?

Third, are we *efficient* in our operation? Are we wasting any resources? Are we being economical? Are we getting the most out of what we have to work with?

Let's look at examples of each. We are on the scene of a commercial structure fire. We only have one ladder truck on scene and no option for another for a long time. We must use this resource safely, effectively, and efficiently. We ask the truck captain to set up on the northeast corner and utilize his aerial to go to the roof for vertical ventilation and be ready for defensive operations. We know it is safe, because he will be out of the collapse zone. To be most effective with our only truck, we have other engine companies perform forcible entry, search, and horizontal ventilation as needed. Since he is in our only truck, we are efficiently setting him up in case we need to go defensive and use elevated master streams, even though we are still offensive. That way, he is in position, and we won't need to move him if things go badly. With only one truck, we can't afford to move him around. The first spot has to be the best for the long haul.

Questions:

- What does SEE stand for again?
- What are all the resources available to me as a captain or BC for my agency? Break them down into the following categories:
  - emergency operations (ops)
  - training
  - discipline
  - purchasing
  - team member assistance
  - outside organizations
  - technical support
  - staffing
  - equipment
  - apparatus
- What fires have I been to in which the resources were not used safely, effectively, or efficiently? What happened?
- What can I do to be more safe, effective, and efficient?
- What can my captain or BC do to be more safe, effective, and efficient?
- What can my fire department do to be more safe, effective, and efficient?

Multitasking skills

This refers to performing numerous tasks and skills simultaneously while monitoring conditions, actions, and needs.

Relevance. In case you are not up on current events, officers must be superheroes at multitasking. Anyone who has seen an officer overwhelmed knows what it looks like. It isn't pretty. We cannot afford to do only one thing at a time. We must be able to SEE through many operations at once.

As a company officer, when you arrive second due to a structure fire, you (your company) should be able to take over command, rapid intervention crew (RIC), and water supply. Although an engineer cannot be part of RIC, the Occupational Safety and Health Administration (OSHA) allows a properly equipped company officer to maintain command and still be part of RIC, as long as you pass command if you must intervene for RIC purposes. This obviously takes skill.

Let's look at a real example. You are the IC of a commercial structure fire. You have heavy fire inside the structure, ventilation is taking place on the roof, and fire attack is inside. No problem, except that you have a total of 12 Mayday calls, and firefighters are down in unknown locations. In the meantime, the fast winds are driving the fire through the concealed spaces of the structure. You must order additional alarms (five total), and evacuate the building. It takes 12 companies to pull out the last firefighter who is not breathing. CPR is initiated and he is transported to the hospital where he is pronounced dead. Word of this travels through your fire scene as you are now into a five-alarm defensive fire with multiple elevated master streams. Sounds like fiction? This is what happened on March 14, 2001, in Phoenix. Do you think the IC was multitasking?

Thoughts. Multitasking takes a lot of practice and time to develop. A great place to start is on the training ground and in the firehouse. Here, lives are not at stake, and you can learn on little things like getting the hose testing done, performing a drill, making a public education appearance, and getting a workout in all before dinner. You will rely on delegation and empowerment to multitask. You must develop trust with your crew(s) through training and time. The ability to constantly perform a conditions, actions, and needs (CAN) analysis is crucial. This will help you multitask.

Let's look a little closer.

- *Conditions* are just that. What is happening? What is the status of the situation? Is it getting better or worse?
- *Actions* involve what is going on. What are you/others doing?
- *Needs* involve what is needed by the people in the conditions who are performing the actions.

For example, you are the IC on a 50-acre grass fire with a rapid rate of spread. You have two alarms on the scene, which equals six grass units, four engines, and two water tenders. You have houses in front of the fire, a creek bed to the west, and a salvage yard to the east. You must protect the homes first and the salvage yard second. You cannot see very well due to the smoke. You have two divisions (sectors) set up and are asking for a CAN check from your divisions. They are your eyes and brains out on the line. Division A states that they have light fuels with minimal fire (conditions), are quickly knocking down the fire along the creek bed on their way to the head (actions), and request that additional grass units go directly to the head to protect the homes (needs). You simultaneously order the balance of the second-alarm grass units to go to the head and order the engines to go to the housing development for structure protection. Division B stated that they have heavy fire in heavy fuels moving into the salvage yard (conditions). They are bumping up to stop it (actions), and require two engines to go to the salvage yard to keep the fire off the fence (needs). You send two engines and one water tender from the first alarm to the salvage yard and one water tender from the second alarm to the middle of the field.

Through multitasking, you were able to do the following:

- define the problems
- set goals and tactical objectives
- prioritize and manage resources in a SEE manner
- do it all in a timely fashion

Questions:

- Do I enjoy multitasking, in fact, *thrive* on it?
- Do I get paralyzed when I have to multitask?
- Do I get the "deer in the headlights" look on my face when a million things are going on around me?
- Do I suffer from analysis paralysis?
- What projects can I do around the firehouse that would require me to multitask?
- How else can I get better at multitasking?

Organizational skills

This refers to creating a systematic and logical way to categorize information, resources, and personnel that enhances effectiveness and efficiency.

Relevance. We must organize our thoughts, information, resources, personnel and operations. As officers, we are charged with getting the most use out of our resources—whether human, fiscal, or physical. As a BC, you may be required to organize staffing in several firehouses. As an IC, you may have to organize companies on the fireground. As a captain, you may have to organize the activities of your crew. The more organized you are in an assessment center, the more likely you will accomplish tasks on time—and do so with little or no wasted effort. This allows you to fine-tune and pay attention to detail, which yields higher scores.

Thoughts. The best way to have an organized work life is to have an organized personal life. Some of us are much more natural at being organized than others. One person may have a pickup truck that is spotless, while the other has a truck that you can't see the floor due to the cups, books, papers, diapers, and receipts scattered everywhere.

A good way to get a handle on this is to have an organizer, a Palm Pilot, or at least a calendar. Start to organize your daily activities, whether personal or work related. At work, you may ask your boss to let you organize a project or a task. For example, you may help your BC with staffing, your captain with training, or the division chief with hose testing. You will learn ways to organize yourself in the assessment center exercises as we discuss each one, but do not limit yourself to those tips. Take the time to develop this skill.

Questions:

- Is my vehicle a pigsty or immaculate?
- Am I always running behind?
- Am I stressed or do I have a good handle on my personal life?
- How could I organize my personal life better?
- What projects or people can I associate with that would help me get better organized?
- Would an organizer or Palm Pilot help?

Planning skills

This refers to planning, organizing, implementing, controlling, and evaluating actions and systems.

Relevance. We must plan both in the emergency and nonemergency environment. Many of our examples thus far all showed the process of planning. Almost all of the components we have discussed overlap when we formulate and implement a plan.

Thoughts. Planning requires time. Another saying to add to our list of quotes is, "If you fail to plan, you plan to fail." This statement is very true. The reason we have *pre-plans* in the fire service is due to the small amount of time we have once a problem (emergency) arises. SOP/SOGs are plans. Once again, they are preexisting in order to be utilized when an emergency arises.

As officers, we plan our shift, plan for training, plan goals for our company (companies), plan for career support of our troops (hopefully), and plan our careers. In the emergency scene environment, we still plan through SOPs/SOGs, pre-plans, strategies, tactical objectives, and tasks.

If you are not a planner now, you had better become one as an aspiring officer. You will need to plan your crews' activities, meals, training, apparatus/equipment maintenance, inspections, workouts, station maintenance, public education events, and special projects. That's your job. As a BC, you can add staffing, budgets, meetings, and reports to the list.

The best way to plan is to start by having a well-organized and well-planned life. Yet another saying: "Make your plans in pencil, because God has an eraser." We must know when to be flexible and when to be firm. Unforeseen circumstances are the name of our business, so things will change. But having a starting point helps tremendously to maintain effectiveness and efficiency.

Meetings are a great way to start the shift. Even if you are interrupted by calls, you can still get some good information to the troops before it all breaks loose. At busy stations, it's common to start a meeting at 8 a.m. and finish it at noon over lunch. That's okay. Just communicate the plan. As a BC, conference calls and e-mails work great to get all of your companies on the same page.

The components of an effective plan are

- time
- goals/objectives
- organization
- communication
- implementation
- monitoring
- adjustment
- contingencies
- evaluation

Let's look at each of these more closely. *Time* must be taken into consideration. How much time do we have? What is the time frame of our plan and when do we want to accomplish the goals? What are the *goals* for the time frame? What *objectives* must be accomplished to complete the goal? We must *organize* the plan to be safe, effective and efficient. We must organize resources, personnel, and perform a strengths, weaknesses, opportunities, and threats (SWOT) analysis. SWOT analysis looks at our environment, both internally and externally.

Next, we *communicate* our plan to the appropriate players. Communication is most often poorly done. Here's another gem: "The biggest problem with communication is assuming it took place." After the plan is communicated, we *implement*. After implementation, we *monitor* the results to see if the plan is working or if we need to adjust. We must have *contingency* plans ready in case our original plan is ineffective. Finally, we *evaluate* the whole process to learn and improve for the next time.

Let's now look at an example of going through the planning process. An incident management team is on a fire and must come up with an incident action plan (IAP) for the next operational period. They have identified the symptoms, problems, alternatives, and solutions. Now they are going to put it all together into an IAP for the next day.

- **Time:** The operational period for the plan is 0600–1800.
- **Goals/objectives:** The goal is to protect life and property and extinguish the fire. The objectives will be cutting hand lines, placing units for structure protection, and deploying air operations.
- **Organized:** The plan is built with SEE operations based on a CAN analysis of their resources and a SWOT analysis of the environment. CAN yields improved conditions, completion of hand lines, and need for additional strike teams to Branch 2. SWOT yielded the *strength* of rehabbed crews and *weakness* of inexperienced crews internally; and *opportunity* for high humidity and *threat* of high winds externally.
- **Communication:** After the plan was developed, it was relayed to the troops prior to start of the operational period during the morning briefing.
- **Implementation:** Crews deployed to their assigned divisions.
- **Monitor:** Safety officers and division supervisors SEE what was going on and if the plan was working with CAN reports.
- **Adjustment:** Operations reassigns more engines to the opposite side of a canyon for structure protection.
- **Contingency:** All troops had the plan of lookouts, communications, escape routes, and safety zones (LCES) in case things went south.
- **Evaluation:** Finally, after the operational period ended and crews returned safely to camp, the planning section evaluated the plan and started on the next operational period IAP.

This is exactly what happens on a large-scale wild land incident, each and every operational period.

Questions:

- Am I a good planner? Or do I fly by the seat of my pants?
- Do I carry an organizer or a Palm Pilot?
- How can I help my captain or BC plan and organize the activities of the company or battalion?
- Where can I find an IAP?
- Did I attend a morning briefing at the last strike team I was on?
- Do we have morning meetings or messages at my company/battalion?
- What else can I do to be a better planner?

Emergency Operations

Although we are going into our third dimension, *emergency operations*, most of the KSAs we have discussed thus far are also required on the emergency scene. The dimensions overlap all the time in the real world, and they also overlap in the assessment center. Notice figure 4–3, where the captain is leading his firefighter, managing information, and performing both on the emergency scene. You will need time management and problem solving on the fireground and during a counseling session. We will look more in-depth at this after we finish this last dimension.

Fig. 4–3 Leadership (people) and management (things) come together on the fireground. Notice the officer multitasking here.

Oral communication skills

This refers to speaking with clarity and being calm and confident on the emergency scene.

Relevance. Although we discussed oral communication in the leadership dimension, we must discuss it again in the emergency operations dimension. Once again, you will use many different KSAs on the fireground and during simulations. Communication is not the only KSA that is used in multiple areas. *During simulations, you* must *verbalize your thoughts (reason for calling another alarm) as well as your actions (apparatus placement, hoseline location, naming of the incident).* We list it again to show another side of communication on the fireground.

Thoughts. We have three Cs that are required on the fireground or any other emergency scene: clarity, confidence, and calmness. As officers, you must provide clear direction to your troops, whether face-to-face or on the radio. If you do not clearly state your orders, they will never come to fruition. Next, confidence must come across in your directions. The troops will not act if they sense fear, uncertainty, or doubt of any kind. Finally, remain calm. If you panic, so will the entire team. If the IC loses his/her cool, everyone else will also since the crew members may be thinking, *The IC must know something we don't.* We will discuss confidence and calmness further later.

Questions:

- Do I get too excited on the fireground? Or do I remain calm?
- Do I give clear direction?
- Is my boss clear, confident, and calm on the fireground?
- How do I feel when responding to a call where the IC sounds freaked out before I arrive?

Ability to remain confident

This refers to providing stability through confidence in adversity.

Relevance. Soldiers don't follow a committee into battle. Firefighters want strong, confident, decisive leadership when times get tough. As an officer, your ability to remain confident during adversity will give your troops stability. Your confidence will transfer to them.

Thoughts. Confidence comes with time. We must continually perfect our craft. *Confidence is the cart that follows the horses of time, education, experience, and trial and error.* You cannot fake confidence. Confidence must sustain the adversity of questions, doubts, and changing conditions. If you are confident in a course of action, take it. But if you have doubt—or those little hairs on the back of your neck start to stand up—stop and *listen* to your doubts. Confidence does not override common sense or prevent you from changing strategy. Just don't change due to peer pressure. Even changing strategy requires confidence. *How* you change strategy is as important as the fact that you did.

Questions:

- Am I naturally confident around my peers?
- Do I need to shore up my knowledge and skills so that I can stand behind my decisions?
- Do I crumble at the first sign of doubt or critical comment?
- Do I know the difference between confidence and cockiness? (See chapter 9.)
- How will I improve my confidence?

Ability to remain calm

This refers to providing stability through remaining calm during adversity.

Relevance. The fireground is no place for out-of-control emotions. Great officers remain calm on the radio and in their face-to-face communications. The world could be burning, but you would never know it if a good calm officer is at the helm. Loosing one's cool will transfer to the troops. (Are we starting to see a theme?) This also applies to the firehouse. No one likes living with an emotional roller coaster for 24 hours a day.

Thoughts. Think of a duck. Although he may look like he is effortlessly moving across the surface of the water, his legs are pumping away. In like fashion, a good officer may have much on his mind: going defensive, calling additional alarms, dealing with irate owners, even a Mayday call from inside. No matter what, the flood of thoughts on the inside (below the surface) must be minimally transmitted (above the surface).

As we discussed previously, stability in an unstable environment is a crucial skill on the fireground. *It is okay, normal, and healthy to have a small dose of fear at times. This kind of fear can keep us safe and out of a bad situation when we listen to our instincts. Panic, however, is* uncontrolled *fear that can lead to chaos.* Remember, there are no emergencies for well-trained firefighters—just problems to solve. Emergencies are for the public. That's why they call us.

Questions:

- Do I get overly excited on calls?
- Am I being honest about the previous question?
- What is the difference between *fear* and *panic*?
- What can I do to improve if needed?

Strategic knowledge

This refers to establishing offensive, marginal, and defensive strategies as appropriate in a SEE manner.

Relevance. As we discussed before, you must be able to plan on the emergency scene. The plan starts at the strategic level and works down through tactical and task levels. The strategic level determines the overall, large-scale plan. As an IC, you must be able to quickly determine (SEE) what strategy to employ. If you cannot do so quickly and accurately, people can die. It's that simple.

Thoughts. Here is the Phoenix Fire Department strategy: *Risk a lot to save savable lives (offensive); Risk a little to save savable property (marginal); Risk nothing to save lives or property already lost (defensive).*

Three strategic modes exist on the fireground: offensive, marginal, and defensive.

- *Offensive* involves interior attack, search, ventilation, salvage, overhaul, etc. You must ensure (SEE) that the window of opportunity to perform is still open. You must justify in real life and during a simulation why it is okay to have people inside.
- The next level is *marginal.* The building could change at any moment, and troops are ready to get out. As the IC, you are in an offensive mode but have notified all companies that we are riding the fence and to be prepared to get out if conditions do not improve.
- Finally, *defensive* is purely exterior. We have cut our losses and are protecting unburned structures. The old Lloyd Layman concept of facts, probabilities, own situation, decision, plan of operation (FPODP) is excellent for determining what strategy to employ.

Questions:

- What conditions would allow an offensive strategy?
- What conditions would require a defensive strategy?
- What would a marginal strategy look like?
- Have I ever been first on the scene of a fire?
- Have I ever been the IC?
- What would require me to take command vs. pass command?
- How is this different on a simulation?

Tactical skills

This refers to employing truck ops, engine ops, special ops, and emergency medical services (EMS) safely, effectively, and efficiently.

Relevance. You must be able to transform strategic goals into tactical objectives. You absolutely must have an excellent working knowledge of engine, truck, EMS, and special ops. You must also know the fundamentals of vehicle accidents, grass fires, HAZMAT incidents, and high-rise fires.

Look at the target hazards in your jurisdiction and ask yourself, "Could I handle being first in and taking command of this?" You may not need to be a confined-space operations-certified guru, but you better know about what resources to call if you are first into a confined-space rescue.

Thoughts. You must be intimately familiar with the fundamentals of basic tactics and strategy.

Let's look at a checklist:

- Size-up
- Command
- Hose line stretching and placement
- Water supply
- Interior attack
- Rescue
- Search
- RIC
- Utility control
- Ventilation
- Forcible entry
- Extrication
- Defensive operations
- Master streams
- Salvage and overhaul
- Wild land
- Strike teams
- HAZMAT
- Multi-casualty incidents (MCI)
- High rise
- Vehicle accidents

Questions:

- What tactical areas do I need to review?
- Am I comfortable being first in to all of these types of incidents?
- How can I become proficient in these areas?
- Are there classes, conferences, workshops I can take?
- Can my current boss (captain/BC) help?

Safety knowledge

This refers to utilizing protective clothing, safety equipment, and procedures safely, effectively, and efficiently.

Relevance. Safety is always the first priority. Remember SEE? Safety must be factored into everything you do. The welfare of your crew depends on it. *As an officer, your main goal in life must be that your crew goes home safe at end of shift.*

Thoughts. Safety must be second nature. You must know when something does not look right. You must develop instincts based on time, experience, and training. You must be the guardian angel of your troops and see the big picture while watching their back. In addition, you should know your department, state, and federal laws and regulations that affect safety. You should have a good understanding of OSHA, two-in-and-two-out, RIC, firefighter survival and many other safety-related issues facing today's firefighters.

Questions:

- Am I safe? Or am I reckless?
- Have I had a greater-than-average number of injuries or incidents?
- What can I do to be safer and then instill that in others?
- What are the fundamentals of command safety?
- What classes and books are out there?
- Are there any other statutes, laws, regulations, policies, or tactics that I should know to be safer?

SOP/SOG knowledge

This refers to knowing and utilizing SOP/SOGs safely, effectively, and efficiently.

Relevance. SOP/SOGs are the playbook for your fire department's emergency operations. You must know them intimately. They will be tested in an assessment center. Even if the assessors do not have an intimate understanding of your department's specifics, the more you are able to articulate them in a simulation, the better you will score. Once again, focus on the job.

Thoughts. Find your SOP/SOGs, know where they are, and practice them. You must know initial first-in procedures, what to do with smoke showing, no sign of fire, options for command, standard alarm assignments, tactical channel usage, etc.

Questions:

- Does my fire department have SOP/SOGs?
- Where do I find them?
- Which ones do I need to practice or review?
- Could I ask my boss if I can sit in the seat as the captain or BC for part of the shift to improve my knowledge and experience base?

ICS/IMS knowledge

This refers to knowing and utilizing components of ICS/IMS safely, effectively, and efficiently.

Relevance. You must be an expert at ICS as it applies to your SOPs, tactics, strategy, and safety. ICS will help you organize your incident from simple to more complex. ICS knowledge will also be required of you on strike teams and interfacing with other agencies. Whether you use the ICS, IMS, or another form of incident command in your area, know it.

Thoughts. ICS will help you organize everything from house fires, grass fires, multi-casualty incidents and HAZMAT incidents to planned events of any kind. As you respond to more large-scale incidents, you will be required to understand the larger system. Review IAPs whenever you have a chance. The myriad of ICS forms will help you understand how the system works. In addition, field operations guides like Firefighting Resources of California Organized for Potential Emergencies (FIRESCOPE) and Fireline Handbooks are excellent resources.

Questions:

- Can I fill out an ICS/IMS chart for my agency/region?
- What is the difference between groups, divisions, and sectors?
- How many branches are in logistics?
- What are the components of the planning section?
- Are units higher than sections?
- Who is in the command and general staffs?
- How and where can I find out all of this?

Ability to exude command presence

This refers to inducing action, maintaining control, and instilling confidence on the emergency scene.

Relevance. Command presence exudes confidence in the troops. You must have the troops follow you to get the job done, but you must be able to take control early. Like herding cats, gaining control over firefighters takes effort and skill. Remember, firefighters want strong leadership. Command presence is simply initiative on the fireground.

Thoughts. As we discussed in the initiative section, few people are willing to step up and take charge when necessary. As an officer, this is just part of the job. This should not be a matter of choice but an instinct to act. Develop this through acting officer work in the position to which you aspire as well as through classes and training. Remember, all the components of command will give you the confidence to have appropriate command presence.

Questions:

- Who do I know with command presence?
- What does command presence look like?
- Why do I want to follow it?
- Do I have it?
- Where else can I get it?
- Will it come overnight?

Assessment Center Exercises and Key Points 5

Okay, we have made it through the foundation and most important part of this book: the KSAs to be an excellent officer. Remember, you may have some and you may need to build some. To find out where you are—and consequently where you want to be—we can utilize some actual assessment center exercises. As in the assessment center, we will look at what KSAs each exercise measures.

As we stated earlier, although the dimensions are broken down into three main categories in this curriculum, multidimensional KSAs can be evaluated in the same exercise. For example, you will need management and leadership skills on the fireground and need to remain confident and calm regarding personnel issues. To better understand the exercises and how the KSAs are measured, we will describe the most commonly used exercises and then list the associated KSAs.

After we look at some examples, we will further explore the exercises by showing sample exercises and evaluation sheets for each exercise type.

Each of the exercises is broken down into the following three areas:

- A brief description of the exercise.
- A list of the KSAs that the exercise evaluates.
- Key points and developing your skills (those particular KSAs).

Remember, it takes planning, dedication, and time to change behavior that will improve a test score and create a great fire officer.

Emergency Scene Simulator

Obviously, the emergency scene simulator is an integral part of a fire department assessment center for a suppression position (lieutenant, captain, BC, etc.) and may comprise the majority of the score. The simulations have evolved over the years. Several styles are used, including the following:

- Overhead picture
- Tabletop picture
- Drawings
- Computer simulations
- Video based

Figure 5–1 shows a typical computer software simulation using a laptop and projector. These have simulated fire and smoke conditions that are extremely realistic and can be changed as the exercise progresses. Multiple sides of the structure can be seen as you make your way around the building.

Fig. 5–1 Modern simulation exercises involve computer-generated scenarios. Digital Combustion, Inc. (www.digitalcombustion.com) has state-of-the-art software and is used by many agencies.

Figure 5–2 shows a simple tabletop picture that is one dimensional. These are less dynamic but easy to practice. One way to practice is to utilize the cover of fire publications like *Fire Engineering Magazine.* Just analyze the cover picture, imagine you are the incident commander, and ask yourself what tactical decisions and priorities are needed.

Fig. 5–2 Although not as sophisticated as computer software simulations, tabletop pictures are excellent tools for practice.

Two main categories of simulations are static and dynamic. In a *static* simulation, you will not have interaction. Both exercises have unique challenges and benefits. In a *dynamic* simulation, you may have interaction with assessors playing the part of companies, dispatch, etc. Radios may be used to enhance realism.

In a static simulation, you will not have radio traffic to cue you. The challenge is that you must verbalize all of your actions or you will not be scored. Most firefighters (FF) like the familiarity of the radio traffic and they become calm. Without the radio traffic, many candidates speak in *radio-ese.* They speak as though they are talking on a radio when there isn't one in the room. Consequently, the assessors do not know the reasons for your actions, and you may assume they know what you are thinking. *If you think it but don't say it, then you don't get credit for it.*

For example, you may say, "Engine 1, on your arrival pull an attack line, you will be interior division." While that's fine, you never told us what size attack line, where you want it placed, and what the tactical objectives are (see fig. 5–3).

Fig. 5–3 You must verbalize all of your thoughts and actions; otherwise you will leave points on the floor.

Think of the static simulation as an oral presentation on how you would fight the fire. Leave no information to chance. Strategy, tactical priorities (and why), location of water supply, reason for calling a second alarm, size and location of hoselines, location and type of ventilation, and a host of other factors must be verbalized to receive credit. Do not mistake the verbalizing of your thoughts as micromanaging. You are simply assuring the assessors that you are thinking ahead and have a complete grasp of the situation.

The benefit of a static simulation is that you do not have the distraction of the radio traffic. For some people, it's easier without radio chatter; for others, the chatter has a calming effect and helps with the realism.

In the dynamic simulation, the challenge is that, in addition to your need to verbalize your thoughts and reasons for your actions, you must manage the radio traffic. Managing both the radio and verbalizing your thoughts can be challenging and takes practice. In the real world, we think a lot on the fireground without needing to verbalize. *In the assessment center, you must think out loud and talk on the radio (see fig. 5–4).*

Fig. 5–4 Having a radio does not relieve you from verbalizing your thoughts.

You also may or may not be required to write a report of your actions after the simulation, so note-taking can be critical.

A plot plan or map of the area may also be made available. Use these to give you a better idea of how to stage apparatus, where to direct hoselines, what hydrants to use, etc. Verbalize and mark on the plot plans/maps so the assessors truly understand your tactics.

The style and media that are used are irrelevant. If you are prepared and have the KSAs described previously, you will do well. *The most important thing to remember is to verbalize all thoughts and actions.* The assessors cannot read your mind. You must verbalize why you are doing what you are doing and the thoughts behind your actions. You should verbalize your initial size-up, crew directions, radio report, and directions to incoming companies. If you call additional alarms, state why.

Always state your strategy and tactical objectives with priorities.

Regardless of the type of simulation, if you do not verbalize, you will not be scored on what you did not say. You will shortchange yourself and leave points on the floor. Once again, these thoughts are based on actual performance of an effective officer. What should you really do?

KSAs evaluated:

- Problem-solving skills
- Planning skills
- Goal-setting skills
- Resource management skills
- Time management skills
- Prioritization skills
- Oral communication skills
- Ability to initiate
- Delegation skills
- Multitasking skills
- Remain confident
- Remain calm
- Strategy knowledge
- Tactical skills
- Safety knowledge
- SOPs knowledge
- ICS knowledge
- Command presence

Key points and developing your skills. Here is a checklist of some main issues to verbalize (and do in the real world). We break them down into four key areas of an incident: dispatch/response, arrival, ongoing operations, and conclusion.

Dispatch/response:

- Confirm address on map
- Route to scene
- Day, time, and weather
- Tactical channel
- Hydrant locations
- Access
- Pre-plans
- Pre-assignment (personnel and companies)
- What companies are responding?
- Is the BC responding?
- Special hazards associated with this occupancy
- Alarm assignment (additional companies based upon updated information)

Upon arrival:

- Size-up (FPODP + 7 sides – *Do a lap and check the rear!*)
- Crew assignments (they are on scene first with you; give them direction or you may lose them)
- Apparatus placement (drive past the building to see 3 sides and out of the "Hot Zone")
- Initial radio report object, conditions, actions, assignments (OCAA):
  - Object – "Engine 1 arrived at 123 10th Street: two-story house."
  - Conditions – "Heavy fire from the second-floor windows."
  - Actions – "Engine 1 will be 10th Street command and has a water supply. We are stretching a hoseline to the front door. Command post is located across the street from the building."
  - Assignment – "Engine 2, stage past E1 and take my firefighter to assume fire attack with the hose off E1. Have your other firefighters pull a back up line."

Operations:

- Safe/effective/efficient
- Strategy (offensive/marginal/defensive)
- Tactical objectives of rescue, exposures, confinement, extinguishment, overhaul, ventilation, and salvage (RECEOVS)
- SOP usage
- RIC/accountability
- Obtain an "All Clear"
- Perform a personnel accountability report (PAR) regularly
- Additional assignments (using ICS/IMS) to accomplish tactical objectives (Truck 1–ventilation group, E3 – search group, Medic 1 – medical group, etc.)
- Secure utilities

- Additional resources:
  - Alarms/apparatus (engines, trucks, medics) (*Stage additional alarms if you do not have an assignment. Assign a staging manager.)*
  - Special ops (light/air units, HAZMAT, rescue, helicopter, etc.)
  - Chief officers (safety, public information officer (PIO), liaison, etc.)
  - Police department/sheriff/highway patrol
  - Other agencies (Fish & Game, county HAZMAT, public works, coroner, etc.)
  - Investigator
  - Utilities
  - Red Cross
  - Station coverage
  - Board-up crew
  - Rehab

Conclusion:

- All personnel safe and accounted for
- Companies back in service
- Turned over to investigator/owner
- Post incident analysis
- Close out

Let's review a checklist of some other fundamentals of tactics and strategy:

- Assume, pass, transfer, or unify command. (For a test, you usually assume command unless otherwise directed.)
- Confirm the address upon arrival.
- If you take command, name the incident and identify the command post location.
- Be aggressive about safety and accountability.
- Have a contingency plan.
- Keep RIC engaged.
- Place apparatus out of the hot zone (collapse, HAZMAT, smoke, upwind, uphill, power lines, etc.).
- Rescue is most important, but hoselines in place may be the first priority to accomplish that goal.
- Communicate your strategy and tactical objectives early.
- Organize and manage the incident with ICS/IMS from the beginning.
- FF safety + RECEOVS.
- Ensure adequate water supply.
- Call additional resources before you need them—plan ahead!
- Give your crew direction before your initial radio report to keep control and get them moving with purpose (prevents freelancing).
- See as many sides as possible—seven (four sides, top, bottom, inside).
- Size up continuously.
- Place hoselines between fire and victims/uninvolved area.

- Use proper size attack lines for fire load and building size.
- Give truck company the front of the building if needed. (Hoses are easier to extend than ladders.)
- Do not ladder over windows.
- Ventilate as directly over the fire as *safely* possible.
- Keep the span of control manageable.
- SEE the big picture.

Simulations can be done in any firehouse. You do not need to set up an extravagant simulation program. Again, just take the cover of any *Fire Engineering* magazine and use that as your scenario. Have officers you respect put you through the paces. Multiple assessors will give you a better cross section of feedback. Practice verbalizing everything that you think and articulate any actions you would take.

Work from simple exercises to more complex ones. Start without radios or maps. Just use a pad of paper, a pencil, and a picture. Utilize the mock exercises in this book or create variations. As you get more complex in your practice sessions, be sure to include radios, maps, pre-plans, plot plans, etc. Try as many different styles of exercise as possible until you are comfortable with anything they throw at you.

Classes include any number of fire marshal command series classes, conferences, tactics and strategy seminars or local drills. Books and videos are available through most department training divisions. Search the Internet under "firefighter training" to find a host of options. In addition, many tactics and strategy books from experts from throughout the country are available. Read magazines like *Fire Engineering.* They each have excellent articles on tactics and strategy. Attend Fire Department Instructors Conference (FDIC) hands-on training (HOT) Sessions.

Drill with your company. Perform acting work on drills. Go around your first-due area and practice radio reports. Talk about size-up with a respected officer. Conduct research and lead a drill on tactics and strategy. Look at target hazards in your area and ask, "What would I do if this thing was burning?"

The Web site www.firefighterclosecalls.com is an excellent resource for information on tactics, strategy, safety, and incident lessons from throughout the country.

Whenever performing a simulation exercise, the first step is often the most difficult. Where do you start? The following pages have some acronyms and tools for getting yourself organized during a simulation. Try them out. If they work for you, great. If not, try something that does. Everyone is different.

Tactics and Strategy Fundamentals
Some tools to get you started—SAW CSS RECEO VSS

The acronym SAW CSS RECEO VSS (defined in the following text) can be used to get you started during the actual simulation exercise. Utilizing an acronym that you have practiced will help you during your simulation. You will have a template from which to set up a checklist. Having a checklist will prevent you from forgetting anything and leaving points on the floor. This checklist contains all the essential elements of tactics and strategy that must be addressed *and verbalized* during a simulation exercise. This has been used in assessment center simulations with great results.

If you have your own acronym or system, then utilize whatever is most comfortable. However, make sure that all of the elements noted as follows are reflected. Once again, do not leave points behind. We will first explain this system and then show how to use it during an actual simulation—when your nerves are competing with your brain.

To remember this acronym, simply practice saying it over and over. Once you study the components, you will easily memorize it since it's a derivative of RECEOVS, which has been around for years. Even if you have never heard of any of this before, with some practice, experience, and common sense, it will come naturally.

You simply must practice your system to have outstanding performance in an assessment center. If you like, you can add department specifics to this if necessary. For the purposes of this book, we will stick to SAW CSS RECEO VSS.

S – Size-up

A – Apparatus placement

W – Water supply

C – Command

S – Strategy

S – Safety

R – Rescue

E – Exposures

C – Confinement

E – Extinguishment

O – Overhaul

V – Ventilation

S – Salvage

S - Support

Let's look at SAW CSS RECEO VSS in more depth.

Size-up: Size-up is the mental evaluation of the situation, meaning FPODP.

This must be performed first in order to set up your plan properly. This must also be done very quickly but accurately as possible with limited information. Size-up is continual. *You must verbalize that you would perform a lap of the building or assign another company to give you a report on conditions from the rear. Don't forget the rear/lap. Many candidates forget to verbalize this critical step.*

FPODP:

- Facts – Large two-story house with fire showing from the first floor at 2 a.m. with one car in the driveway.
- Probabilities – Could have a victim inside above the fire.
- Own situation – One truck and two engines are arriving.
- Decision – We will use an offensive strategy.
- Plan of action – Engine 2 will be fire attack, Engine 3 will be rescue group, and Truck 1 will be ventilation group. We will place an attack line on the fire between the fire and the occupants, send a search team upstairs, ladder the building, and ventilate.

Apparatus placement: Verbalize that, if safely possible, you would drive past the structure to see three sides and get as much information as available. Before you step out of the cab/driver's seat, you will have seen three sides of the problem.

Out of the hot zone:

- Collapse
- Heat
- HAZMAT
- Upwind/uphill
- Power lines

Water supply: You must be able to support your operation.

- Who (Which company has it?)
- What (What is the type of supply? Hydrant, tender, etc.)
- Where (Where is the source?)

Command:

- Who (Are you taking it or passing it?)
- What (What is the name?)
- Where (What is the command post location?)

Strategy: Verbalize if it is offensive, defensive, or marginal.

Sets the general direction of the operation

Safety: Once the strategy is set, you can then take appropriate safety measures. Safety measures for offensive versus defensive strategies are different (personal protective equipment [PPE], crews inside, etc.).

Key points to verbalize:

- RIC
- PAR
- Fireground accountability tracking systems (FATS)
- Utility control

Rescue: The first tactical priority is rescue, although hoselines in place may be required to accomplish this objective.

Key points to verbalize:

- Primary search complete
- Secondary search complete
- All clear

Each of these should be done by different crews to ensure a fresh set of eyes has completed each task. Plan ahead by having treatment/transport capabilities (ambulances) on scene before you need them.

Exposures: Exposures must be addressed both internally and externally so that you see the big picture without tunnel visioning on the seat of the fire.

- Internal (attic/cockloft, floor above, other rooms, egress, stairwells, etc.)
- External (all four sides of the occupancy, downwind, adjacent occupancies, common attics, etc.)

Confinement: A critical item to verbalize is your hoseline placement. This shows the assessors where you are initiating your attack, where the fire is being pushed, and what area you are defending in the structure. Remember, protect the uninvolved areas, occupants, and means of egress.

Extinguishment: As a compliment to confinement, extinguishment gives you the opportunity to state what caliber hoseline you have chosen. Do not bring a knife to a gun fight. We can sometimes make the mistake of using bread and butter house fire tactics on commercial occupancies. Big mistake! Don't get into the habit of pulling small caliber weapons to big battles.

Overhaul: Verbalize who is performing the overhaul and in what locations. For example, in a balloon-frame building, you would check extension in the walls and attic appropriately.

Ventilation: Obviously, ventilation is a key point. *Hit it early and aggressively. Many candidates forget or wait too long to verbalize ventilation.* You must verbalize the location and nature of the ventilation. Is it vertical, horizontal, pressurized, offensive (heat cuts), defensive (trench cuts), etc? This must be done in concert with interior crews, which requires good, clear communication.

Salvage: Along with verbalizing the performance of salvage operations, state your priorities. For example, a two-story apartment building with fire on the second floor would require the first floor to be a priority salvage location due to potential water damage. In addition, state that you would contact the owner/occupant and that you would support their needs as necessary.

Support: Additional resources may be needed to support the operation and must be called ahead of time. Stay ahead of the power curve. If you get behind, it's hard to catch up. *Anticipate your needs before they arise!*

For example, many candidates wait to call an ambulance until they have a firefighter down or a victim located. That's way too late! If you have a working fire, call an ambulance. Verbalize that they are there to "provide any required treatment and transport of potentially injured victims or firefighters." This shows that you are thinking ahead, being proactive rather that reactive.

Resources to consider calling ahead of the need:

- Alarms (engines, trucks, ambulances)
- Personnel (safety officer, public information officer, technical specialists)
- Special equipment (air unit, HAZMAT, rescue)
- Other agencies (law, utility companies, Red Cross, Fish & Game, etc.)

A key point here is to set up another acronym of the resources you would likely call in your jurisdiction. Every agency is different and has different resources available to them. For example, if you know you would call a safety officer, an investigator, and law enforcement, then "SIL" could be placed in this section of your worksheet. Again, the goal is to have a usable checklist that will keep you from forgetting anything during the assessment center simulation exercise.

Size-up vs. arrival report

Now that we covered the basics of SAW CSS RECEO VSS for the test, let's look at the difference between a size-up and an arrival report.

Two different things: Size-up happens *before* the arrival report.

- Size-up: *The mental process of gathering information about the situation (FPODP) in order to make a decision. Size-up is continual.*
- Arrival report: *The first unit's radio transmission that gives a situation report, states the plan of action, directs incoming resources, and orders additional resources (OCAA). The initial arrival report happens once (see fig. 5–5).*

Fig. 5–5 Do not mistake a size-up for an initial arrival report. Your initial arrival report happens once, while the mental size-up is continual.

Arrival report:

- Object – Two-story house.
- Condition – Fire on the first floor with occupants trapped on the second.
- Action – Engine 1 taking Elm Street Command on the "A" side and stretching a line to the front door for fire attack.
- Assignment – Engine 2 assume rescue group. Engine 3, get a water supply and pull a back up line. Truck 1 assume ventilation group and assist with rescue.

How do I use this in the test?

You will be given the dispatch information either verbally or in writing. You will most likely have 2–10 minutes to prepare for the exercise after you are given the information. This preparation time will allow you to get yourself organized. You can set up a tactical worksheet using SAW CSS RECEO VSS to keep track of your resources.

You must have a way to condense the tactical fundamentals previously discussed into a usable format during the test. Typically, you will have blank sheets of paper and pencils to use during your simulation. During your preparation time before the exercise, write the acronym SAW CSS RECEO VSS down the left side of the paper, as shown in figure 5–6. While that may seem like a mouthful, with a little practice writing it out, it's easy to remember since it is based on tactical fundamentals. This will give you a self-made tactical worksheet to use during the simulation so you do not forget anything.

Fig. 5–6 The SAW CSS RECEO VSS acronym will get you started. You should spend considerable time perfecting your worksheet. You must be able to write it out quickly.

Once again, if you have your own system to create a tactical worksheet, use it. This one is an option or you can modify it for your department.

As you write these tactical objectives down, you are creating a tactical worksheet, as seen in figure 5–7.

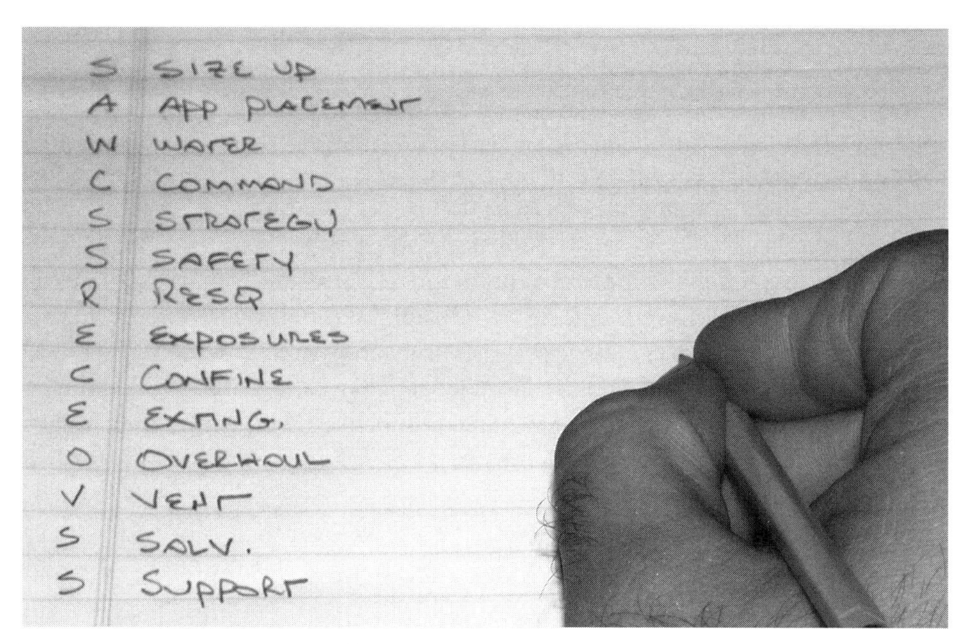

Fig. 5–7 Next, SAW CSS RECEO VSS should be written out. Although a lot to remember, it's easy with practice.

In addition, you must elaborate on your tactical worksheet in order not to forget anything and be scored appropriately. Each component of SAW CSS RECEO VSS has key points (discussed previously) that will be evaluated by the assessors.

Let's look at how to condense the key points for the simulation exercise.

S – Size-up:

- ❑ FPODP
- ❑ Seven sides
- ❑ Lap

A – Apparatus placement:

- ❑ Past 3 sides
- ❑ Out of hot zone

W – Water supply:

- ❑ Who
- ❑ Where
- ❑ What

C – Command

- ❑ Name
- ❑ Incident command post (ICP)

S – Strategy

- ❑ Offensive
- ❑ Defensive
- ❑ Marginal

S – Safety

- ❑ RIC
- ❑ PAR
- ❑ Utilities

R – Rescue

- ❑ Primary search
- ❑ Secondary search
- ❑ All clear

E – Exposures

- ❑ Internal
- ❑ External

C – Confinement

- ❑ Hoseline location

E – Extinguishment

- ❑ Hoseline size

O – Overhaul

- ❑ Who
- ❑ Where

V – Ventilation

- ❑ Horizontal
- ❑ Vertical
- ❑ Location

S – Salvage

- ❑ Who
- ❑ Where
- ❑ Reporting party (RP) contact

S – Support

- ❑ Safety officer
- ❑ Investigator
- ❑ Law
- ❑ Air
- ❑ Utility company
- ❑ Red Cross
- ❑ Rehab unit

Looking at figure 5–8, you can abbreviate these components even further to minimize the time it will take you to write out your tactical worksheet checklist. Remember, this will take practice to get written out. At test time, it should only take you about a minute to write the abbreviated checklist out (as shown in fig. 5–8). Although the exercise directions may give you up to 10 minutes or more to prepare and set up your worksheet, practice for one minute so that you are ready for a much shorter time frame.

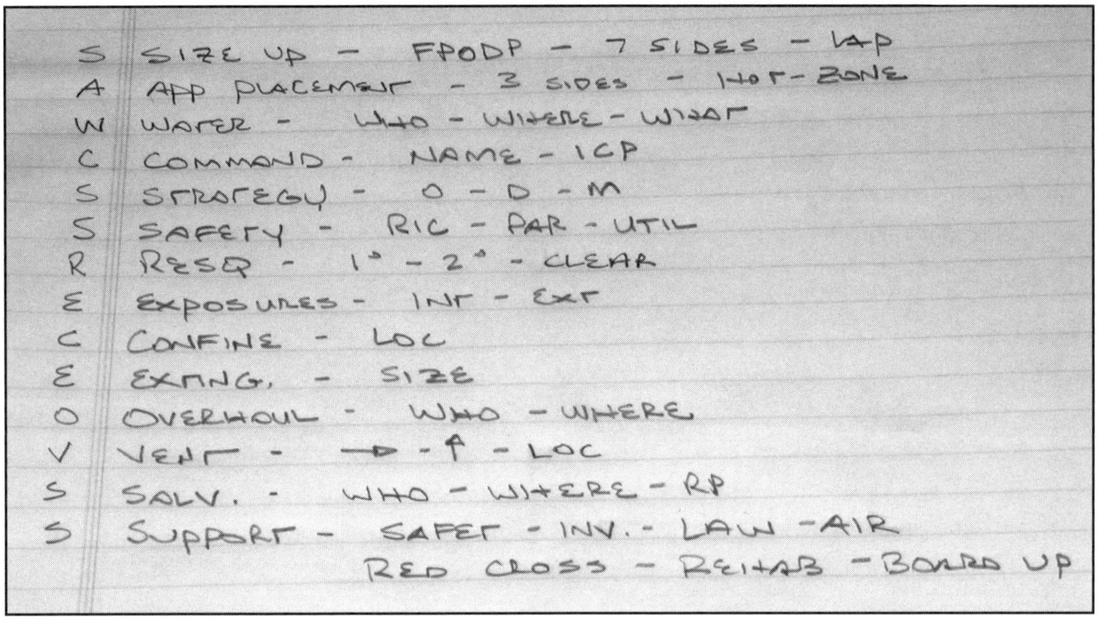

S SIZE UP - FPODP - 7 SIDES - LAP
A APP PLACEMENT - 3 SIDES - HOT-ZONE
W WATER - WHO - WHERE - WHAT
C COMMAND - NAME - ICP
S STRATEGY - O - D - M
S SAFETY - RIC - PAR - UTIL
R RESQ - 1° - 2° - CLEAR
E EXPOSURES - INT - EXT
C CONFINE - LOC
E EXTING. - SIZE
O OVERHAUL - WHO - WHERE
V VENT - → - ↑ - LOC
S SALV. - WHO - WHERE - RP
S SUPPORT - SAFET - INV. - LAW - AIR
RED CROSS - REHAB - BOARD UP

Fig. 5–8 Each SAW CSS RECEOVSS component has additional key points. By making a checklist, you will not leave points on the floor.

Remember, you should set up your own list of support resources that you would call. Practice writing out the resources you would call to support your plan, as shown in figure 5–9. Write them out in the order you would typically request them from dispatch.

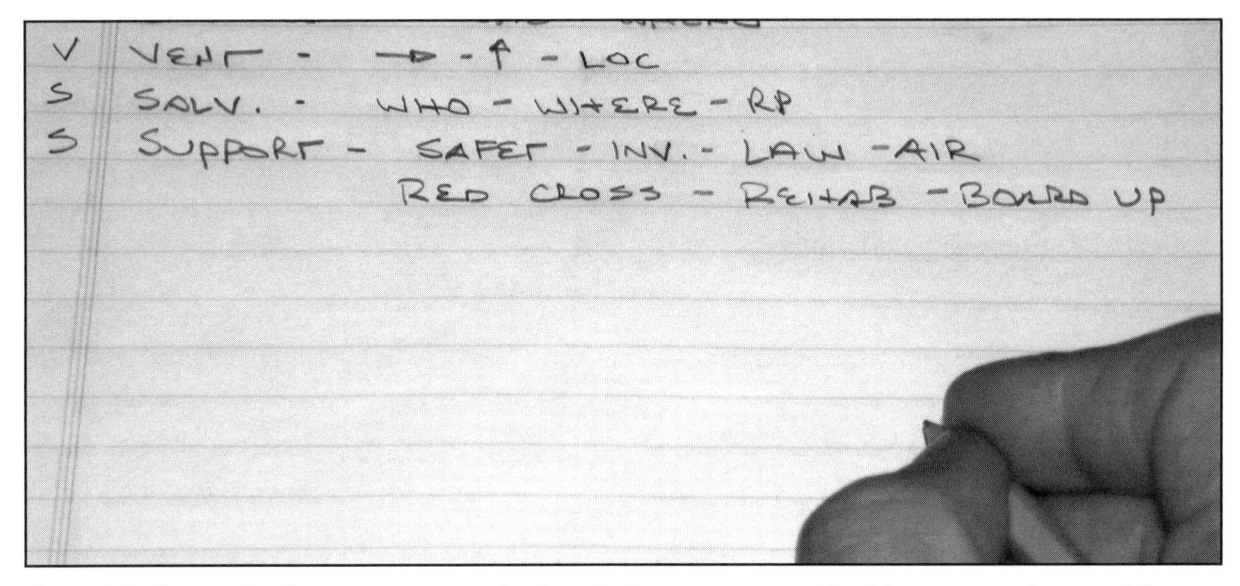

Fig. 5–9 For "support," make your own acronym for the typical resources you would call in your agency to support the operation.

Keeping track of your resources

Keeping track of your resources is critical. Safety, effectiveness, and efficiency all rely upon it. Assessors may ask you where your companies are located and what their tactical objectives are. To keep track of your resources, utilize your local ICS/IMS at the very beginning and throughout your scenario. This will help you keep a manageable span of control.

Using ICS for example, rather than assigning Engine 1 to perform fire attack and continuing to call them "Engine 1," call them "Interior Division" or "Fire Attack Group" as appropriate. You can then assign subsequent companies to them as needed. Engine 2 would then report to "Interior Division." You will keep a more manageable span of control as your incident progresses.

Let's look at a commercial building fire as an example. If you have four engines, two trucks, and a medic company responding, you have seven companies on the scene. Engines 1, 2, and 3, and Trucks 1 and 2, and Medic 1 all report directly to you. You are at the limit of your manageable span of control if each company reports directly to you and you call them by their company name.

By using ICS, you could have an interior division comprised of three engines and one truck with the tactical objectives of extinguishment and rescue. The second truck could be your ventilation group, and Engine 4 and medic could be RIC with the tactical objectives of performing a lap, securing utilities, and performing fireground accountability. Using ICS, your span of control went from 7:1 to 3:1, as shown in figure 5–10.

When you write down a division/group/sector name (Interior, for example), the first company listed under that name should be the division/group supervisor. Looking at figure 5–10, Interior Division is supervised by Engine 1, Ventilation Group is supervised by Truck 2 and RIC is supervised by Engine 3. This will help you keep track of those specific companies that have been assigned to supervise and report directly back to you. During a simulation, you may be asked who is responsible for a particular division. This will give you a quick reference.

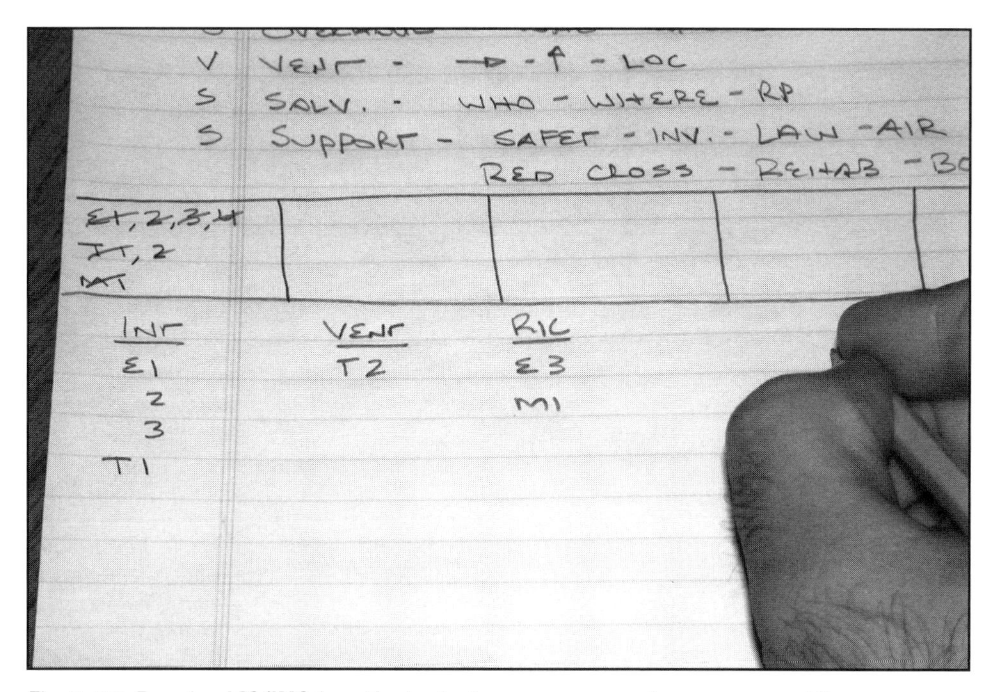

Fig. 5–10 By using ICS/IMS from the beginning, you can organize your scene while demonstrating your ability to use the system properly. List the division/group supervisor companies first under each respective group/division.

Let's say you are concerned about the adjacent occupancy igniting and involvement in the attic of the original building. Now, if you called a second alarm, you could have an exposure group with the tactical objective of protecting the second building and send one of the second alarm trucks to the roof of the first building to assist the ventilation group as needed. Although you doubled your alarm assignment to 14 companies, your span of control would only grow from 3:1 to 4:1, as shown in figure 5–11. Interior division, ventilation group, RIC, and exposure group all report to you, representing 14 companies.

When the assessors give you the companies on your second and third alarms, just write them down in an area designated on your worksheet. As you assign them to their respective groups or divisions, cross them off as shown in figure 5–12. If they are not crossed off, consider them in staging.

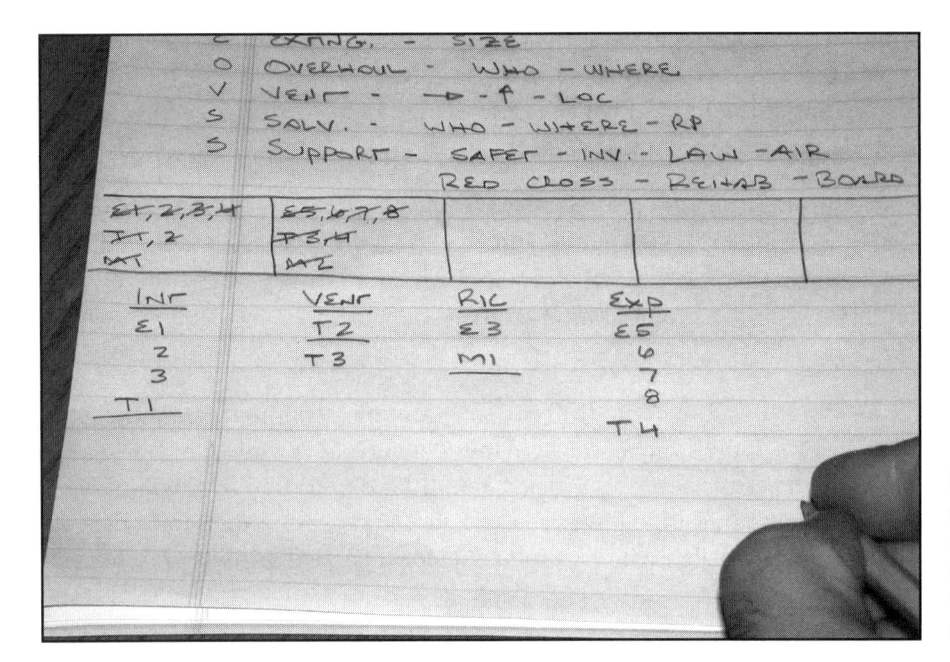

Fig. 5–11 Although we added seven companies, we have only expanded our span of control by one. Simply place the companies where you want them in the system.

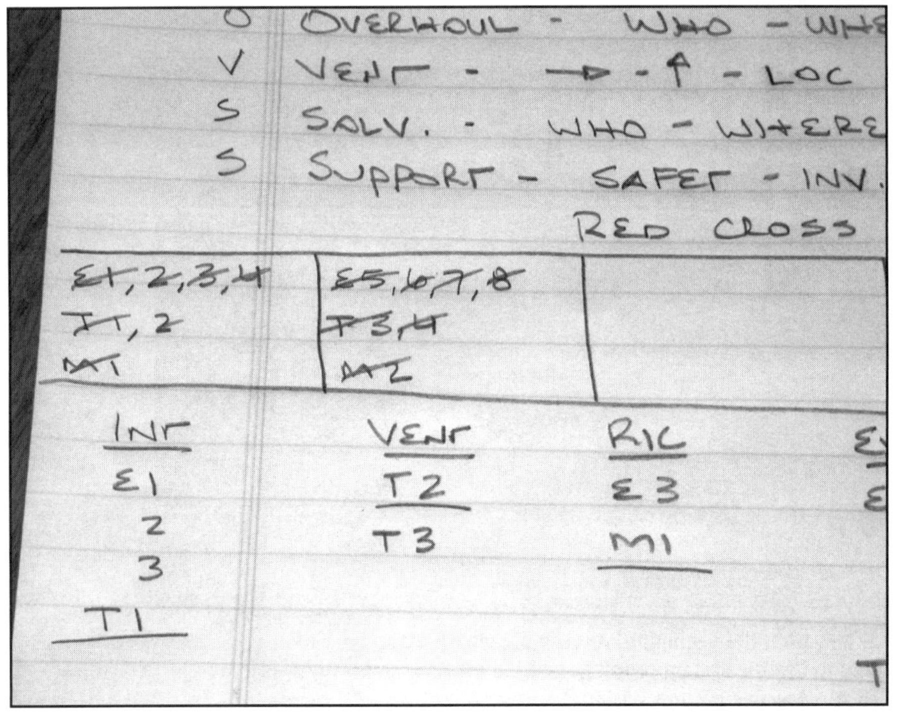

Fig. 5–12 Each box on your worksheet represents an alarm. The second box has the second alarm, etc. As you place companies into the system, cross them off. If they are not crossed off, consider them in staging.

What do I say and when?

Make sure you understand the rules of the exercise before you begin. Ask the proctor or assessors questions if needed. Clarify if the exercise begins upon dispatch or upon arrival. If upon dispatch, then verbalize according to the "upon dispatch/response" section we discussed earlier (looking at the map, finding hydrants, etc.). Some candidates will not make this distinction and leave points on the floor by not verbalizing critical response information they would assess or actions they would take while responding.

If the exercise does not begin until the picture comes up, just wait, get your tactical worksheet organized, and focus.

When the picture first comes up on the screen, you will be "on scene." At this time, most candidates make the mistake of immediately giving an arrival report without taking adequate time to perform a mental size-up. Due to nerves, they start speaking immediately in radio-ese, that dreaded language of the nervous and ill-prepared candidate.

You will be smarter than that. When the picture first comes up, say, "Now that I am on scene, I would perform the following quick mental size-up." This shows the assessors that you are smart enough to size up the situation before you act. Since you are telling them that this is your *mental* size-up, you are thinking out loud and racking up the points.

Simply go through the FPODP of your picture. Let's say you are the BC responding to a structure fire with two engines and one truck. For figure 5–13, the mental size-up might sound like this: "Now that I am on scene, I would perform the following quick mental size-up. **Facts**, I am on scene of a 1-story house with fire on the A/B corner. **Probabilities**, this could be occupied due to the truck in the driveway. My **own situation** is that I am on scene alone with two engines and one truck responding. My **decision** is going to be an offensive strategy. My **plan of action** is to have the first engine attack the fire from the A-side front door and push the fire out the A/B corner. The truck will ventilate vertically."

About now, the assessors are feeling inspired and see that you know your stuff. They are eager to hear more. You then say, "That being said, I would give the following arrival report to the incoming units on the radio. Battalion 1 arrived; one-story house with fire on the A/B corner. Battalion 1 is assuming Main Street IC. First arriving engine, you will be interior division, your tactical objectives are fire attack from the A side and search."

Now, the assessors want to pin a badge on your chest. You have smoothly made a mental FPODP size-up out loud, then given an OCAA arrival report. You are off to a great start!

From there, simply continue to verbalize your thoughts, reasoning, and actions and track your resources on your worksheet. Continue to verbalize things like the command post location, your lap around the structure to see all the sides, and company tactical objectives, and run down your SAW CSS RECEO VSS checklist so you do not miss anything. Check off components as you go so you are sure you covered everything and scored the maximum points, as shown in figure 5–14.

Fig. 5–13 Practice verbalizing the FPODP size-up with pictures or when driving in your first-due area.

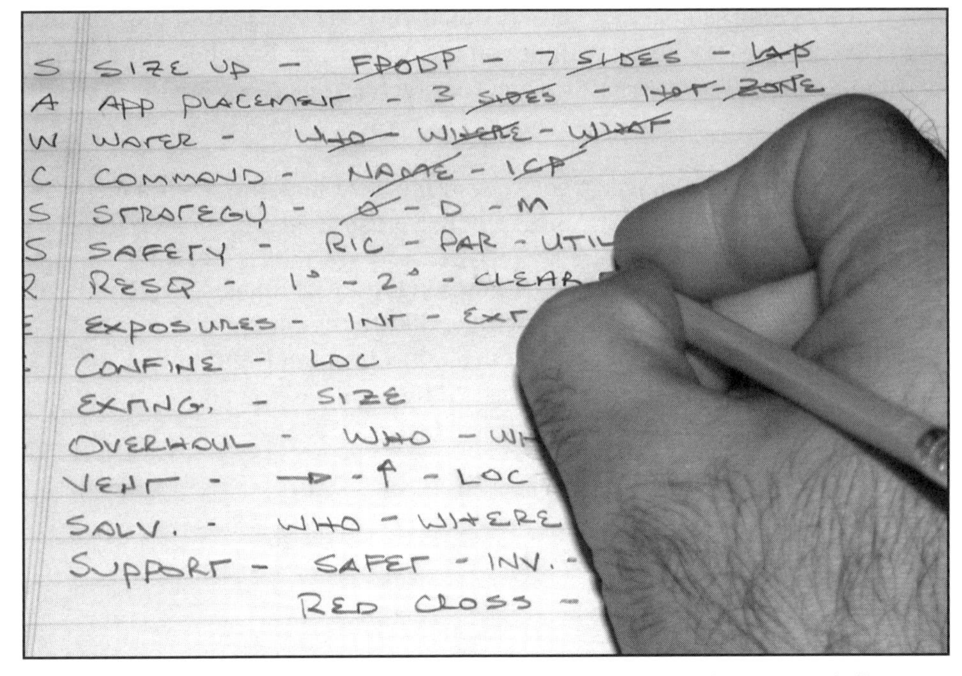

Fig. 5–14 Check off the tactical priorities as you go, ensuring that you have covered all your bases.

Time management in the simulation

Ensure that you are aware of the time that you have to prepare and the time to mitigate the problem. As an example, you may have 10 minutes to prepare, 2 minutes to respond, 10 minutes to mitigate the problem, and 2 minutes to conduct a pass on to the chief officer taking over for you. All of these times are realistic in an assessment center. *The key is to read the directions and underline the key time components.* Set your stopwatch on your wristwatch to 00:00 and be ready to start it when the time comes. You want your watch ticking on the same time as the assessors' clock.

Remember the importance of training and repetitive practice in the simulation. As we said in the beginning of the book, *at the moment of truth, whether in a test or during the real thing, you will not rise to the level of expectation, you will fall to the level of your training.*

In-Basket Exercise

The in-basket exercise is one of the oldest and most widely used exercises in assessment centers, whether in the fire service or private sector. In-baskets are designed to evaluate your overall decision making, ability to prioritize, written skills, and problem solving. The specific KSAs are listed as follows. You will be given a finite amount of time to go through a pile of memos, e-mails, or phone messages. You must prioritize the pile and deal with the issues. You may be asked to write a response to each issue, stating what action you would take. You may also be required to actually write any correspondence (memo, e-mail, and letter) that is appropriate to the action you take. The key here is to follow the directions, prioritize, and use time effectively.

The first step is to look at the time, and ensure you know when you need to be done. Write that time down and circle it somewhere so that you can be easily reminded.

Second, look at the size of the exercise. Like a quick size-up, skim over the pages to see the amount of issues, memos, e-mails, etc. This will help you understand the overall complexity and scope of the exercise. You can then pace yourself accordingly.

The third step is to read the directions twice. The first time, read through them to get an overview. The second time, read slower and underline/circle/highlight critical pieces of information. For example, you would underline, "Today is August 1, and you are the captain of Engine 10. You will be on vacation for 2 weeks starting tomorrow." This gives you a proper perspective from which to act. Another example to underline would be, "Write any e-mails and be prepared to give an oral presentation stating any actions you would take." If you miss this critical step, you will not receive credit. Plan your time accordingly to allow for this part of the exercise.

The fourth step is to go through the entire pile of communications. An effective strategy is to place each item into one of three piles: immediate, delayed, and minor (just like triage), as shown in figure 5–15. Even though some items may seem so incredibly immediate that they may urge you to act, *do not act until you have gone through the entire pile.*

Fig. 5–15 Sort the in-basket items in three priority piles.

The reasons for this are to (1) see the big picture, (2) delete any communications that are redundant, and (3) locate any changed or canceled communications. For example, the top memo may state, "Call the Fire Chief ASAP about an issue with your engine at the grocery store." That may panic you to act immediately. Do not. The seventh memo dated a day later may state, "Cancel call to the Chief, it was not your shift. Sorry for the confusion."

Another tool is to number each item in the top right corner in the pile of correspondence as you go so you can refer to them more easily and attach related items when necessary, as shown in figure 5–16. You are not numbering in priority order, just use sequential order so you can refer to related issues later. Clip related items together or put the number of the related item on the bottom right corner of the page of the opposing item, as seen in figure 5–17.

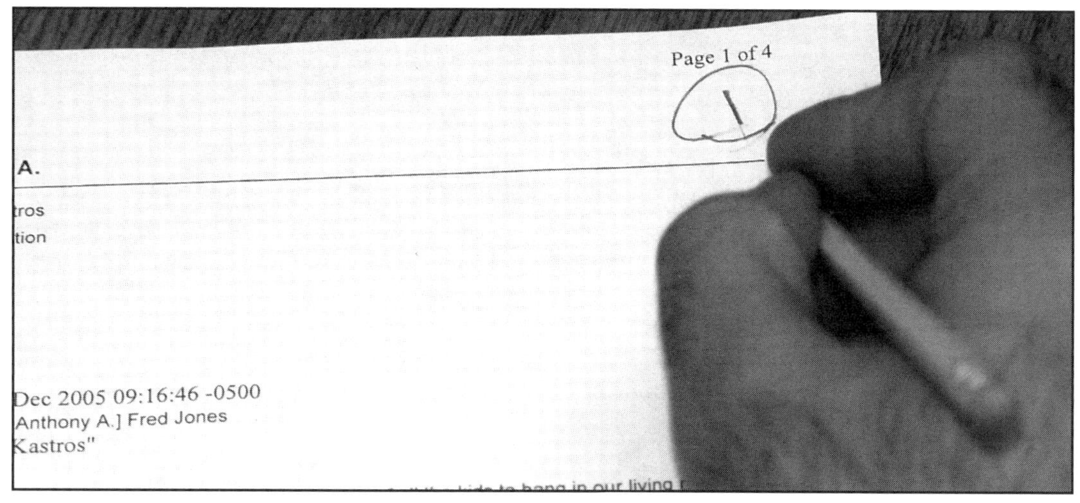

Fig. 5–16 By numbering the items, you can keep better track of related items later.

Fig. 5–17 Clip or fold over the corner of related items so they remain attached. You can also place the number of the related item on the bottom right corner.

After you go through the memos and categorize them, start acting on the *immediate* pile of issues first, then the *delayed*, and then the *minor* issues last. Look for trends, related issues or contradictory memos (like the one noted previously). You may actually havc to act on only half of the ones that you started with (thus saving valuable time), and some related issues may be in different piles.

Other trend examples would be a memo stating that you must be at a drill today at 1 p.m., but another memo states the school teacher at the elementary school down the street is looking to book you at 1:30 p.m. since the other shift kept blowing her off.

Which is the priority? The priority in this case would be the elementary school if that's consistent with your department values and culture. If we care about customer service, then the fact that the school teacher has been blown off should raise a red flag in your mind. Look at the option of combining these two issues. If the drill in question is a wet drill, perhaps you can conduct the drill at the school so the kids can watch. The company gets to drill and the kids get to see you in action.

In general, fire service priorities are (1) safety, (2) emergency response, and (3) training/customer service. Training and customer service vary from department to department. One fire department may not hesitate to take a company out of service for a customer service event, while another may cringe at the thought of it. It depends upon your own department culture, leadership philosophy, and priorities. Know them and you will act appropriately. You must have a good knowledge of your policies so that you act accordingly and go up the chain of command when appropriate. Remember, bosses don't like surprises—but don't go to your boss to ask permission to wipe your nose either. Keep them in the loop or pass up the chain as appropriate.

Manage yourself to act within the time allowed. Budget your time with planning, prioritization, and organization. You will probably be required to give a presentation and/or answer questions on the actions you took. Also be prepared to write appropriate correspondence. Be prepared to articulate the reasons for your actions and communications. *Bottom line: follow the directions.*

KSAs evaluated:

- Written communication skills
- Oral communication skills
- Delegation skills
- Problem-solving skills
- Policy knowledge
- Time management skills
- Prioritization skills
- Multitasking skills
- Planning skills

Key points and developing your skills. Learn your key policies. Know where you can find them. Ask your boss to let you take care of running the shift for a day (see fig. 5–18). Address any messages, phone calls, e-mails, meetings, drills, etc. Decide what the day's activities will be. Look at the big picture. Check for scheduled drills, maintenance, public education events, etc. You must have the support of your boss to make this work. Ensure that your boss' boss knows about it as well. If you are aspiring to be a BC, while you are off duty, ride along with one you respect. Find out the tempo of the day. See what issues face him/her in a day and how they are dealt with.

Fig. 5–18 This candidate is practicing the job at the firehouse before the test.

Modified In-Basket

The modified in-basket is a derivative of the traditional in-basket. The modified in-basket is best described as a "day in the life of" a particular position. A list of events is given, usually in chronological order. Rather than place the items in separate piles, you must decide how to handle each event in order. After some time to go over the list and develop a course of actions, you will be asked to give an oral presentation of your actions and answer appropriate questions. This requires an intimate knowledge of policy, SOPs, and an ability to initiate action appropriate to an event.

Here is an example of some items that may be listed in a modified in-basket.

- 07:45 – Engineer calls in sick.
- 08:03 – Off-going firefighter wants to talk to you about his captain.
- 08:19 – Emergency call, medical aid.
- 08:52 – Return to station. Callback engineer is at station.
- 09:15 – Callback engineer states that rear flashing lights do not work on the engine.
- 10:10 – BC calls and wants to conduct a drill at 13:00.
- 10:30 – E-mail from Mrs. Smith from Silly Elementary School. She wants to know if you will bring the engine by at 13:30 since B shift blew her off twice.
- 11:30 – Firefighter burns himself with scalding hot water while making lunch for the crew.

These require many of the same skills as an in-basket but are presented in multiple forms of communication and events.

KSAs evaluated:

- Oral communication skills (see a pattern?)
- Ability to initiate
- Delegation skills
- Problem-solving skills
- Policy knowledge
- Goal-setting skills
- Time management skills
- Prioritization skills
- Resource management skills
- Multitasking skills
- Planning skills
- SOP knowledge

Key points and developing your skills. This is similar to the traditional in-basket, but you also must know your available resources. Know your authority level. What are you able to do with/without permission and from whom?

Do not give simple answers to complex events. For example, the 0803 item where the firefighter wants to talk to you about his captain could be addressed by saying, "I would see what he wanted and act appropriately." The more in-depth answer would be, "I would assess whether this was a sensitive issue like harassment or something simple like a surprise party for the captain. If the issue is sensitive, I would ensure that we have privacy, document, and act accordingly. I would also ask the firefighter if he approached the captain about the issue since this is an across-shift issue. I would want the opportunity to solve a problem with a member of my crew, so I would extend that courtesy to this captain if possible. That being said, something like harassment would require me to involve the BC, and the human resources division, according to policy. Once again, I would act differently based on the situation. The key is to listen, assess, process, and act accordingly." See the difference? The top candidate will go the extra mile and give an in-depth response.

Written Exercise

The written exercise evaluates many KSAs besides simple written communication. The KSAs are listed as follows. Some keys to effective writing will also be discussed.

Typical assessment center written exercises deal with issues currently facing the department (morale, budgets, hiring, training, policies, communications, harassment, etc.). Another common exercise is to have the candidate write a report/narrative of the emergency simulation exercise performed previously in the assessment center. You may be asked to give an oral presentation and/or answer questions about your paper.

Regardless of the type of document, you must budget your time during an assessment written exercise. Although some chief-level assessments will allow you to take the exercise home due to the complexity of the issues, most company officer-level written exercises will require you to write on the spot.

KSAs evaluated:

- Written communication skills
- Delegation skills
- Problem-solving skills
- Policy knowledge
- Goal-setting skills
- Time management skills
- Planning skills
- Emergency operations (for report narratives)

Key points and developing your skills. First, know how to write. That may sound simple, but unfortunately many firefighters do not know how to write. We are so action-oriented that we lose whatever skills we may have had in high school or college. We must know how to organize our thoughts in an effective and brief format. Let's look at some keys to effective writing.

Writing takes education and practice. Content *and* format are important to a good composition. You must be educated on the effective methods of writing and then practice by writing about the issues facing your department. Ask your boss and staff members about the issues that are going on. Perform a SWOT analysis of the environment.

Hundreds of books have been written on how to write effectively for a myriad of topics and styles. Go to any large bookstore or browse the Internet to find what you need to address your particular weaknesses (general writing, grammar, syntax, sentence structure, outlines, spelling of common words, word usage, etc.).

Many community colleges offer very good writing courses. Once you have become educated, practice writing about your department's issues and ask respected officers to read your compositions.

The most effective way to write, especially in an assessment center, is to start with an outline. An outline will organize your thoughts and prevent wasting precious time. You will want to make a statement, make an observation, or send a message in your paper. This statement or message is called a *thesis statement*, which is the nucleus of your outline.

For example, if asked what the number one problem is in our fire department, your thesis could be, "Ineffective communication is the number one problem facing Dream Fire Department due to emotions, improper media, and insufficient effort."

That *thesis statement* would be supported by *topic sentences*. Each topic sentence would begin a paragraph that supports the topic sentence and thesis.

For example, the first topic sentence could be, "The first issue causing ineffective communications is emotions from past events and relationships." Then your paragraph would support that statement.

After your three topic sentences and supporting paragraphs are completed, you would summarize and conclude. "In summary, ineffective communication is the number one problem facing Dream Fire Department. Emotions, improper media, and insufficient effort hamper effective communications."

Now let's look at an example outline for this document:

I. Introduction
   a. Many problems face Dream Fire Department. In order to properly solve them, we must prioritize the problems in order of significance.
   b. (Support paragraphs)
   c. *Thesis statement* – "Ineffective communication is the No. 1 problem facing Dream Fire Department due to emotions, improper media, and insufficient effort."

II. "The first issue causing ineffective communication is emotion from past events and relationships."
   a. Layoffs.
   b. Contract negotiations.
   c. Labor/management relations.
   d. Potential solution.

III. "The second issue causing ineffective communication is improper media."
   a. Videos.
   b. E-mails.
   c. Solutions.

IV. "The third issue causing poor communication is insufficient effort."
   a. Apathy.
   b. Feeling there isn't a problem.
   c. Solutions.

V. Summary
   a. "In summary, ineffective communications is the No. 1 problem facing Dream Fire Department. Emotions, improper media, and insufficient effort hamper effective communications."
   b. Solution statements.
   c. Closing.

Once again, budget your time by writing an outline in your assessment center. *Do not expect to have enough time to write two drafts of your document.* Write the outline and then write the document *once.* Many candidates have run out of time and turned in half-completed papers, winning another trip to the assessment center two years later.

As stated earlier in the emergency simulation section, you may be required to write an incident report or narrative for your written exercise. Many exercises are related, creating a greater sense of realism.

Oral Presentation

As we have stated, oral communication skills are essential to the fire officer. Like a written exercise, you must organize your thoughts, use time efficiently, and be prepared to answer questions. Many oral exercises are in the form of a teaching demonstration, drill, public education simulation, or oral resume. Be ready for anything. Remember, if you have the skills, it doesn't matter what they throw at you.

Once again, *be the position to which you are aspiring.* (See fig. 5–19.)

Fig. 5–19 Notice the candidate using the easel pad to introduce himself as the officer.

Our candidate is meeting with a community group in this exercise. He has introduced himself as the captain. This sets the tone and shows that he is in the proper mindset even before the presentation begins.

Here are some key points:

- Define your role.
  - Lieutenant and captain – mostly about presentation
  - BC and above – also about content at this level (more depth required)
- Define your audience.
  - Speak to their level in their terms.
  - Make it interesting/motivational for them.
  - What is the goal (inform, educate, solicit, respond)?
- Format your topic.
  - Three sections is ideal (a-b-c, 1-2-3, begin-middle-end)
  - Intro/preview, body, review
- Use your tools.
  - Easels, pens, props, your hat, your badge, the room, the listener— everything is usable (see fig. 5–20).
  - If you have an easel, use it as your notes/outline. Do not try to manage notes and an easel. That's too messy.
- Have fun, be real, connect.
  - Use humor, look them in the eye, be honest.
- Review and allow for questions.
- Stay on time.

Fig. 5–20 Use anything to help in your presentation: a hat, a badge, a patch. This makes the presentation more original and fun.

Be Captain Smith, or Battalion Chief Jones. When you address your audience (the assessors), be the position. If they are playing the part of your crew, city council, senior staff, parents, or otherwise, then treat them as such.

Let's look at an example that happened in an actual assessment center for captain. We have turned it into an exercise here to help prepare you for the assessment.

You are given the task to read a chapter about fire station safety and give a drill to your crew in an hour. In that hour, you review the chapter given to the candidates on station safety and organize your presentation. You have been given an easel pad and colored pens to use as you see fit (see fig. 5–21). A key point here is not to spend too much time reading the chapter. You are being evaluated on your presentation skills, not your ability to recite station safety principles. This is a dry topic that you are challenged with making interesting. Therefore, simply look at the high points of the chapter without burning up time reading every word.

Fig. 5–21 Here, the candidate is writing notes for his presentation. Light pencil notes are for the candidate, while the larger bullet points are for the assessors. Both can be used simultaneously to organize a presentation.

When the time comes, you begin your presentation by saying, "Hey gang, I wanted to have a brief but important drill on firehouse safety. Since Joey burned himself with hot water today (remember that earlier event in the modified in-basket?), I thought we could all benefit from discussing a few safety basics. Tom (you look at one of the assessors), what is the big rule for lifting?"

Tom the assessor goes from assessor to student/crew member. He actually gets a bit nervous since you caught him off guard and he nervously says, "Uh . . . lift with our legs, right?" Figure 5–22 shows the rapport you can develop with assessors.

Fig. 5–22 Try to engage the assessors and get them involved to make it more fun and realistic. If they do not respond, just keep moving along. Don't force it.

"Yeah, great job Tom!" You say. Then you notice the other members of your crew sit a little straighter and take notice because they want to be ready for the next question. Before you know it, they forgot that *they* were the evaluators. *You* are the boss! See how it works?

If they do not give you feedback or play along, don't force it, just keep moving along. Some assessors are instructed not to interact with candidates. That's fine, just be positive, articulate, and confident and have fun. Simply use your questions as more rhetorical to transition to new ideas.

Time management is again critical. The only thing worse than running out of time is not using enough time. If you have 10 minutes to give your presentation, shoot for 9 minutes. Do not leave points on the floor by rushing. Lend depth and insight to your topic. Set your stopwatch and keep track of your time throughout the presentation, as shown in figure 5–23. If you find you have too much time left, slow your pace. If you need to speed up, do so. Either way, watch your watch.

If you have an easel pad at your disposal, use it. Create an outline on the easel pad from which to give your presentation. Do not attempt to manage the easel pad and a handful of notes. This can become difficult to manage well. Everything you need should be on the easel pad. Think of it as a manual PowerPoint presentation. If you want to scribe notes on the easel, write the smaller notes in pencil and write the main key points that you want the assessors to see in large colored ink. They won't see the pencil notes if you write lightly. Another technique is to fold the lower corner of all of your sheets on the easel pad so you can grasp them easier.

In addition, leave a blank sheet in between all of your written sheets on the easel pad (see fig. 5–24). This has two benefits. First, the listeners will not see the bleed-through of the next sheet. Second, if you want to add a sheet/slide after you have begun writing out your presentation, you have the extra blank sheet.

Fig. 5–23 You must keep an eye on your watch during any exercise. During an oral presentation, you can check each time you turn the page or randomly throughout the presentation. Watch your watch!

Fig. 5–24 Leave a spare sheet between presentation slides. If you have a blank sheet after a written sheet, turn both at once to prevent turning extra sheets and having a busy presentation.

Like the written presentation, take some time to brainstorm and put together an outline in your mind. The time you spend up front brainstorming will pay big dividends in your effectiveness.

As described previously, the chief officer candidates cannot simply get by with a nice presentation that has no meat, no substance. As a chief officer, you are expected to have more depth, a greater understanding of the systems in your department, and most importantly, an in-depth plan to fix a problem. Many chief officer tests allow you to take the questions/topics home for research. This allows you to do your homework before the day of the assessment center, quite differently than the fire station safety example given here. You must then show your in-depth knowledge of the problem and solution. Do not forget implementation, evaluation, etc.

A recent example was a BC assessment center where the candidates were given three months to research topics, including, "How to narrow the generation gap and plan for succession." As you can see, this is a bit more detailed than fire station safety.

You must assume that you are expected to have a very specific plan if you were given three months to conduct research. State the situation (symptoms), the problem, the solution, the plan for implementation (step-by-step), and the means to evaluate the effectiveness of the plan, just as you would as a real BC.

Oral/Visual Resume

Another oral presentation variation is the oral/visual resume. This is not as common as it once was because the topic is very subjective. If the goal of the exercise is purely to assess presentation skills, then the oral resume can be distracting to the assessors. Candidates with a great resume may have poor ability to present the content but could be scored higher due to the content itself. Conversely, the oral resume that is designed to see what the candidate has done to prepare himself is a useful tool. Either way, it's good to practice this exercise.

If using an easel pad, resist the temptation to simply make an enlarged resume document. You can be much more effective, creative, and interesting by utilizing a metaphor from which to explain your history. The strategy for the oral resume is very personal. Choose a metaphor that you can relate to most. For example, if you coach or enjoy football, draw a football field to show your progression of training and experience. If you love baseball, utilize a baseball diamond with the crowd as the citizens, the bases as your accomplished career goals, and the players as your co-workers who helped you accomplish your goals (hit you around the bases). The possibilities are only limited by your imagination. Other metaphor examples include ladders, burning buildings, mountains, etc.

One very effective example used by a captain candidate was a drawing of a firefighter. He showed us his firm foundation grounded of family and faith. His legs were EMS and fire, to show a strong footing of balanced experience. His light was vision, planning ahead for the future. His gloved hand was sooty, showing his experience. His breathing apparatus was his motivation to keep going, and his red helmet was his beacon of leadership. With each of these metaphors, he gave solid examples of experience or education to back up his drawing. He scored #1 on this exercise.

Do not limit your resume to the past. Connect your past experiences to your future goals as an officer. Show the panel how your education and experience will be utilized in your new role.

KSAs evaluated:

- Oral communication skills
- Interpersonal skills
- Motivation skills
- Team-building skills
- Time management skills
- Planning skills
- Leadership skills

Key points and developing your skills. Get up in front of people as often as possible. Around the firehouse, lead drills, give presentations to your crew, and lead station tours. When you go to the store, hand out fire prevention coloring books. Lead school presentations, and public education events that your company is assigned to. Ask the training division if they need help teaching classes or conducting multi-company drills. Teach probies or at your fire academy. Those folks are usually more afraid than you are. If you like EMS, become an adjunct instructor for an EMT class to build upon your EMS skills while honing your oral presentation skills.

Many other outside opportunities exist like community college speech classes. Organizations like Toastmasters also can be a great chance to practice.

Chief officer candidates should do all of the steps *and* have an in-depth knowledge of the larger issues facing management. Some topics that you should research include: sick leave use, succession planning, training, budget constraints, expansion of operations, labor contract issues, labor/management relations, cultural boundaries, etc. Have a solution and a plan for problems associated with these issues and practice writing/presenting your solutions/plans.

Role-Play/Counseling Exercise

The role-play/counseling exercise will challenge your interpersonal skills, leadership skills, and problem-solving ability. You will be given information on a crewmember or member of the public with whom you must interact (see fig. 5–25). A problem exists, but you have limited knowledge of the symptoms or their causes. Through interpersonal skills, you will define and solve the problem. Once again, *be the officer.*

Never allow the interaction to get out of control. Never lose your temper or let the other person control the situation. Listen attentively and be respectful, but do not be a pushover.

A key to remember is to be *firm*, *fair*, and *friendly*. You are seeking to solve a problem, but you must motivate the other person to take part in the solution. Once again, *no one cares how much you know until they know how much you care.*

Fig. 5–25 The counseling session will test your people skills and problem-solving ability.

Make the environment safe by placing the chairs on the same side of the table to prevent an adversarial stance, as shown in figure 5–26. Start small. Take time to establish rapport with the person. Ask him/her how life is going. Remind the person of the good things they are doing. Do not open up with comments like, "We have a problem." You will shut down communication, and the person will go on the defensive.

Fig. 5–26 Set the area up for a safe and comfortable environment. Sit on the same side of the table as your role player.

While you are listening, look for the real problem behind the symptoms that you are given. Ask specific questions to identify what is going on behind the symptoms. You may be told that your engineer is always late. Being late is a symptom. You must look past that to find out *why* he is late. Does he have a problem at home or child care issues? Is he going through a divorce? Only when you identify the true problem can you then apply an effective solution.

Look for subtle comments. If you ask how things are going at home, the role player may say, "Home is home." That should cue you to ask further probing questions. Do not gloss over these questions too quickly or you could miss something.

Never take the issue personally, even if they do. Always end with an agreement on a positive note.

KSAs evaluated

- Interpersonal skills
- Oral communication skills
- Motivation skills
- Ability to initiate
- Team-building skills
- Empowerment skills
- Consistency
- Put others first
- Problem-solving skills
- Policy knowledge
- Goal-setting skills
- Time management skills
- Planning skills

Key points and developing your skills.

- Welcome them and establish rapport. (Don't rush this part.)
- Make the environment safe. (Sit on the same side of the table.)
- Give reason for the meeting.
- Separate the problem from the person. (Bad act does *not* always equal bad person.)
- Remind them of the good things they have done.
- Look at trends in symptoms to find the *real problem.* (Late to work = symptom; divorce = problem.)
- Help them open up—motivate them to communicate and take ownership of the situation/problem.

- Remain *firm, fair, and friendly*—remember this is still the workplace, and you have a responsibility as the boss.
- Give *expected behavior*/needed change (policy).
- Establish a plan *together* to solve the problem.
- Help where *you* can/give solutions (coaching, employee assistance, counseling, training, etc.).
- *Agree* on action together (see fig. 5–27).
- Establish *time frame* for behavioral change.
- Establish *monitoring* criteria.
- State *consequences* if behavior does not change.
- End on *positive* note/value of person to organization/you want to help them—"Help me help you."
- Document accordingly.

Fig. 5–27 Agree at the end with a firm handshake and a plan.

This particular exercise is a lot of fun to practice around the firehouse. Have someone portray a disgruntled engineer, a salty firefighter, or an irate citizen. Ask them to really get into the role and make you work to look through the symptoms to find the real underlying problem.

Over the long term, the best way to improve your interpersonal skills and conflict resolution is by asking for honest feedback on where you can improve. Ask trusted peers and bosses where you need some improvement and sincerely work on it. This is much easier said than done. Usually, it takes a significant emotional event to truly change someone. Get involved with group projects in the department and see how well you get along with people. Do people like to work with you? Do you always seem to have "issues" with people?

Look at what motivates behavior of the different people in your department. Watch and listen a lot. Observe and analyze. Do a little psychoanalysis. Look for symptoms of poor behavior and delve into the true problems that cause it. Remember, bad act does not equal bad person.

Find great mentors to emulate. See how they treat others with patience, compassion, empathy, fairness, etc. Learn from them.

Supervisory Exercise

The supervisory exercise involves a series of scenarios in which you must present your diagnosis of the problems and actions taken to solve them. Unlike the in-basket and modified in-basket, the supervisory exercise will give you more detailed events that cause you to formulate a plan and take action (see fig. 5–28).

Fig. 5–28 You will need to quickly but thoroughly analyze each situation in the supervisory exercise.

Usually, you will be given several scenarios dealing primarily with personnel issues or customers. You may have half a dozen issues or more that you must evaluate in as short as 20 minutes. Then, you may be required to give a presentation to a panel of assessors stating your intended actions, as shown in figure 5–29.

Fig. 5–29 Ensure that you elaborate and move at a steady pace during the supervisory presentation phase. If you move too fast, you will not give enough depth to your answers.

As the name of the exercise indicates, this is about supervision and dealing with people.

The KSAs are similar to the subordinate counseling exercise:

- Interpersonal skills
- Oral communication skills
- Motivation skills
- Ability to initiate
- Team-building skills
- Empowerment skills
- Consistency
- Put others first
- Problem-solving skills
- Policy knowledge
- Goal-setting skills
- Time management skills
- Planning skills

Key points and developing your skills. Like the in-basket, you should budget your time and take some time up front to get a quick overview of the problems you are facing. Be sure to read through all the scenarios while not running out of time. Underline key points, times, and directions.

Make notes on each sheet/scenario page that indicate what the problems are, what actions you would take, to whom you would communicate, and what policies are relevant.

This is about problem solving and people. You will be required to sift through the scenarios (symptoms) and articulate your definition of each problem. As you state your solution, identify in detail the steps you would take and how you would document/follow up.

- Preview the entire stack of scenarios to size up the number and complexity of the problems.
- Look at trends in symptoms to find the real problem.
- Identify the people involved.
- Identify any policy or safety issues that must be addressed.
- List the communications you would make and to whom.
- List the actions you would take to solve the problem.
- Identify a plan to follow up and monitor.
- Identify what will happen if conditions do not improve.
- Look for related issues.
- If giving an oral presentation to indicate your actions, ensure that you are prepared.

A great way to prepare for this exercise is to simply practice using mock exercises. Remember, this is a combination of the in-basket and subordinate counseling session. Ask your boss to create scenarios for you that involve people, problems, policies, etc. Ask your boss to write down a series of scenarios, time you, and have you give an oral presentation. Start with one or two in 30 minutes. Then, work toward six in 20 minutes. Increase the complexity as you get more proficient. Real-world incidents that have happened in your department are great mock exercises.

Leaderless Groups

Leaderless groups are utilized less frequently now than they have been in the past. A leaderless group places several candidates in the same room with a common goal that they must attain together. For example, a group of BC candidates may be asked to develop a list of important SOPs to consider for later development. They are told to limit the list to only the top 10 most important procedures.

The group will then be evaluated by several assessors. Ideally one assessor per candidate should be utilized, but this can be very difficult to manage, especially with a large list of candidates.

As a candidate, you do not necessarily need to be in charge. Once again, this is a leaderless group. Effective positions in the group could be one of scribe, clarifying agent, or simply constructive team player.

KSAs evaluated:

- Interpersonal skills
- Oral communication skills
- Ability to initiate
- Team-building skills
- Empowerment skills
- Put others first
- Problem-solving skills
- Policy knowledge
- Goal-setting skills
- Time management skills
- Planning skills

Key points and developing your skills. Take an active role in the process. If you are appointed the leader or chair of the group, then ensure everyone is heard and valued. You do not have to be in charge to be successful. There may not be a formal leader chosen. As long as you participate, contribute, communicate, and work well with everyone, you will do well. You are not competing with the other candidates in the room, only with your own potential.

Listening is just as important as speaking. You must demonstrate your ability to work with others and not dominate the situation. Empowering and valuing others is considered positive. That being said, do not sit back and watch the entire exercise or you will leave points on the floor.

Position yourself to actively participate. Try not to have your back to the assessors. Good assessors will remain mobile or set up the room so that no one is placed in an unfair position. Figure 5–30 shows several people participating in a leaderless group. Notice all are actively engaged, while not any one of them seems to be formally in charge. You could be a scribe, writing down and clarifying the opinions of the group. You could be responsible for a particular aspect of the group, such as generating a written report. Just remain engaged, contribute, and work well. *Forget that it's a test, and focus on the task at hand, like a real team or committee would.* Let your natural abilities speak for themselves. Do not try to force it. Remember, you are not in competition with anyone but yourself.

Fig. 5–30 The leaderless group will test your ability to work with others. Focus on the task, not the test.

Mock Exercises 6

This chapter is comprised of mock exercises so that you can apply what you have learned so far. Utilize some of your respected peers and bosses as assessors. Look for successful role model officers or those who have some assessment center experience. Each exercise has a respective score sheet to further assist you in meeting the associated KSAs. You can photocopy the exercises and score sheets out of the book to have multiple opportunities to practice.

Start slowly. Don't make the exercises too complex at first. Feel free to modify them to better suit your department or simplify them until you get more comfortable. Remember, the key is the KSAs and the best way to determine your level is through practicing the exercises. Ensure that you hit the key points for each as described in chapter 5.

The score sheets for each exercise are generic. They are not specific to any fire department. They are extremely thorough to give you an opportunity to see the types of rating systems used. These are very similar to what assessors would use in an assessment center. The simulations have two different score sheets to give you some variety in the scoring criteria. One is very specific, with check boxes. The other is very subjective, requiring a numeric value from the rater.

Emergency Simulation Exercises

Emergency simulation exercise #1

You will be evaluated on your emergency scene management skills. During this evaluation, you will be scored on your ability to manage and mitigate an emergency problem.

You will be evaluated on the following 5 criteria:

1. Strategy and tactics
2. Use of the ICS
3. Command presence
4. Safety
5. Problem solving

You will be dispatched to an emergency scene. You will arrive three minutes after you are given the dispatch information. Upon arrival, you will be shown a picture of the scene from which to start your exercise.

You will then have up to 10 minutes to manage the problem. After 10 minutes, the assessors will ask questions for up to five minutes. Please verbalize all of your actions in order to be scored appropriately. A pad of paper and pencils are available for use.

Do you have any questions?

You are dispatched to 5150 Elm Street for an unknown type fire. You are the captain of Engine 1. In addition, Engine 2 and Truck 5 are dispatched. You may call additional units as needed. You will arrive in three minutes.

(Fig. 6–1 shows the fire scene simulation window. Fig. 6–2 is the corresponding plot plan.)

Fig. 6–1 Simulation exercise #1.

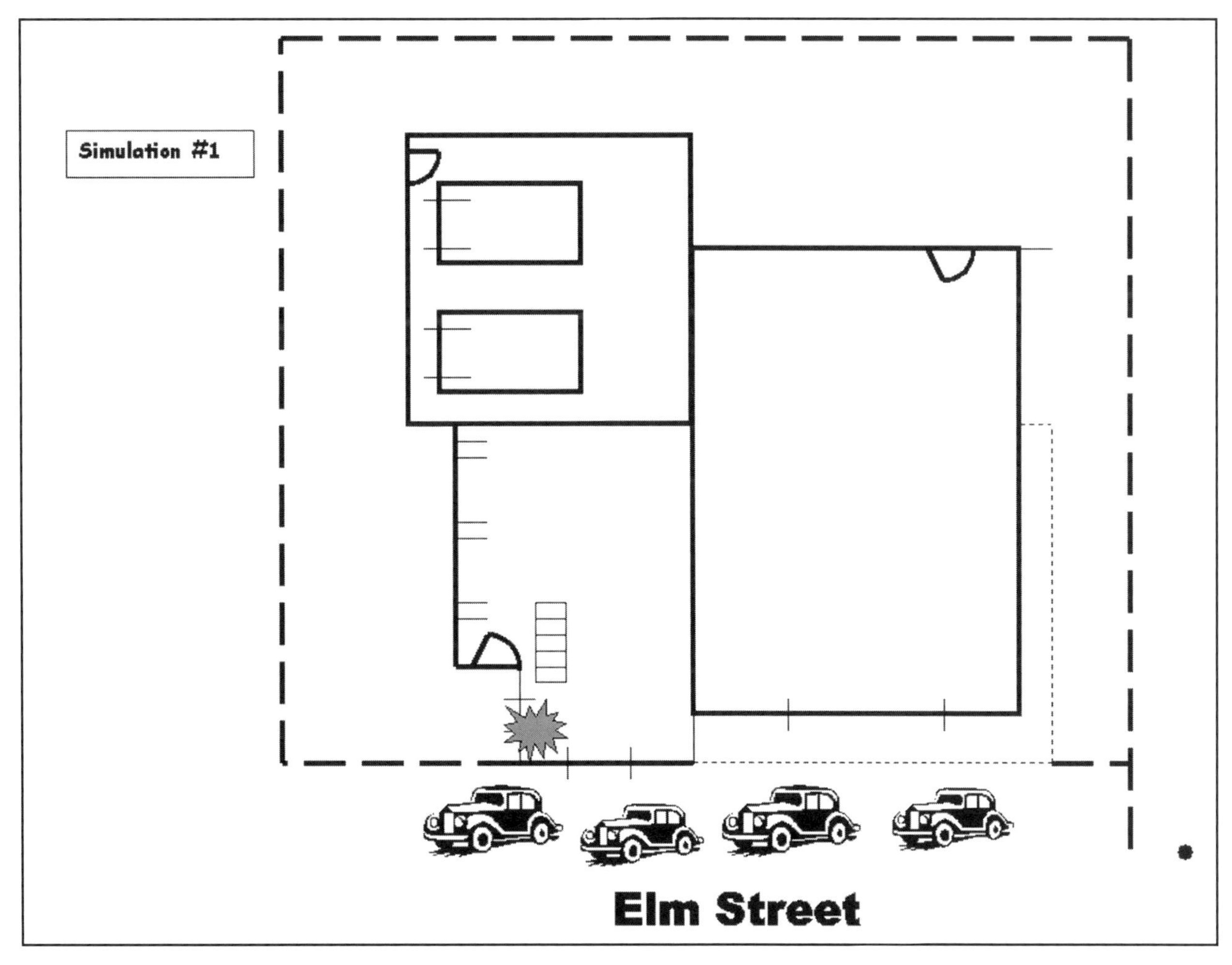

Fig. 6–2 Simulation #1 plot plan.

Emergency Simulation #1 Score Sheet

Candidate Name: ______________________________

Strategy and tactics: (2 points each) **Score:** ________

- Gave initial radio report with appropriate information upon arrival
- Stated strategy for incident (offensive vs. defensive)
- Stated tactical objectives (fire attack, rescue, ventilation, etc.)
- Ordered appropriate resources in a timely fashion to support plan
- Tactical priorities of life safety and property conservation maintained in proper order
- Proper sized hoselines chosen with sufficient water supply established
- Hoselines placed between fire and occupants/uninvolved area
- Safety/RECEOVS components all utilized
- Ventilation established early
- "All clear" for primary and secondary search obtained

Use of ICS: (2 points each) **Score:** ________

- Named incident
- Identified command post location
- Utilized divisions and groups appropriately
- Maintained manageable span of control
- Called additional officers for command staff
- Gave division/group supervisors tactical objectives
- Requested additional tactical channels as needed
- Communication plan established
- Built ICS that was appropriate size and structure for the incident
- Had resources report directly to division/group supervisors

Command presence: (4 points each) **Score:** ________

- Maintained calm demeanor
- Showed confidence in abilities and decisions
- Would inspire action and confidence on scene
- Comfortable in seat of command
- Controlled resources effectively

Safety: (2 points each) **Score: ________**

- All apparatus staged in safe and effective locations
- Personnel accountability maintained
- Safety officer utilized
- RIC established with appropriate resource support
- PAR taken on regular intervals
- Willingness to go to defensive strategy if needed
- Rehab considered
- Medic units requested before being needed
- Identification of special safety concerns verbalized (lines down, collapse, etc.)
- Ensured proper PPE worn by personnel

Problem solving: (2 points each) **Score: ________**

- Looked at symptoms to define root problems
- Attempted to see seven sides of the problem
- Evaluated contingency plans and alternate options
- Developed solution based on problem, not symptoms
- Communicated plan to all personnel
- Saw potential problems at incipient stage, took quick action
- Monitored effectiveness of plan
- Adjusted accordingly
- Planned ahead
- Conducted post incident analysis

Total Score: ________

Comments:

Emergency simulation exercise #2

You are the captain on Engine 7. You will be dispatched to an emergency scene. All of your actions must be verbalized in order to receive credit.

You will be given a note pad and pencil to be used during this exercise. After dispatch, you will have two minutes to arrive on scene. Upon arrival, you will be shown a picture of the scene. You will then have 10 minutes to complete the exercise. The assessors will ask questions after the simulation is completed.

Do you have any questions?

The time is 4 p.m. on Sunday. Engine 7, Engine 1, Engine 3, and Truck 5 are dispatched to a structure fire at 3130 Madison Ave., #4. Battalion 1 is delayed and will not be on scene for 10 minutes. He has asked you to take command. You will arrive in two minutes.

(Fig. 6–3 shows the fire scene simulation window. Fig. 6–4 is the corresponding plot plan.)

Fig. 6–3 Simulation exercise #2.

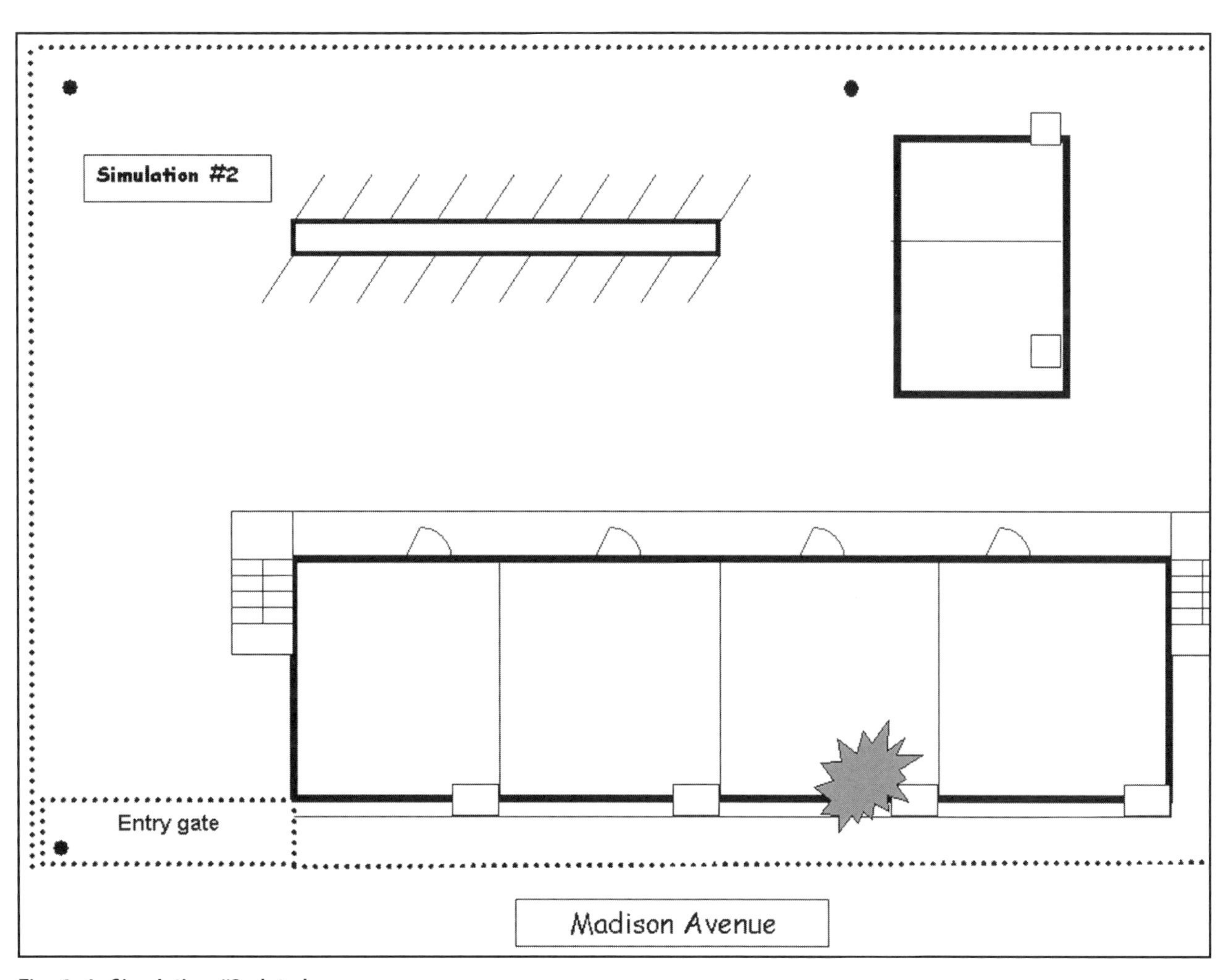

Fig. 6–4 Simulation #2 plot plan.

Emergency Simulation #2 Score Sheet

Candidate Name: ______________________________

Strategy and tactics: (2 points each) **Score:** ______

- Gave initial radio report with appropriate information upon arrival
- Stated strategy for incident (offensive vs. defensive)
- Stated tactical objectives (fire attack, rescue, ventilation, etc.)
- Ordered appropriate resources in a timely fashion to support plan
- Tactical priorities of life safety and property conservation maintained in proper order
- Proper sized hoselines chosen with sufficient water supply established
- Hoselines placed between fire and occupants/uninvolved area
- Safety/RECEO/VS components all utilized
- Ventilation established early
- "All clear" for primary and secondary search obtained

Use of ICS: (2 points each) **Score:** ______

- Named incident
- Identified command post location
- Utilized divisions and groups appropriately
- Maintained manageable span of control
- Called additional officers for command staff
- Gave division/group supervisors tactical objectives
- Requested additional tactical channels as needed
- Communication plan established
- Built ICS that was appropriate size and structure for the incident
- Had resources report directly to division/group supervisors

Command presence: (4 points each) **Score:** ______

- Maintained calm demeanor
- Showed confidence in abilities and decisions
- Would inspire action and confidence on scene
- Comfortable in seat of command
- Controlled resources effectively

Safety: (2 points each) **Score: ______**

- All apparatus staged in safe and effective locations
- Personnel accountability maintained
- Safety officer utilized
- RIC established with appropriate resource support
- PAR taken on regular intervals
- Willingness to go to defensive strategy if needed
- Rehab considered
- Medic units requested before being needed
- Identification of special safety concerns verbalized (lines down, collapse, etc.)
- Ensured proper PPE worn by personnel

Problem solving: (2 points each) **Score: ______**

- Looked at symptoms to define root problems
- Attempted to see seven sides of the problem
- Evaluated contingency plans and alternate options
- Developed solution based on problem, not symptoms
- Communicated plan to all personnel
- Saw potential problems at incipient stage, took quick action
- Monitored effectiveness of plan
- Adjusted accordingly
- Planned ahead
- Conducted post incident analysis

Total Score: ______

Comments:

Emergency simulation exercise #3

You will be required to demonstrate your emergency scene management skills during this exercise. You will be given a written dispatch. After two minutes, you will be on scene. At that time, you will have 10 minutes to complete the emergency scene simulation. You must verbalize all of your actions in order to be scored appropriately. You may call any resources that you would to mitigate this problem.

During your exercise, you will have the use of paper, pencils, colored markers, and a plot plan of the scene. You may use these tools to take notes, clarify your actions, and record your progress. After 10 minutes, the assessors will ask you a series of questions.

You are Battalion 12. You are dispatched with Engine 23, Engine 24, Engine and Rescue 21, and Truck 23 to 6000 Coyle Avenue for a structure fire. The time is Tuesday at 11:00 a.m. You will arrive in two minutes. Do you have any questions?

(Fig. 6–5 shows the fire scene simulation window. Fig. 6–6 is the corresponding plot plan.)

Fig. 6–5 Simulation exercise #3.

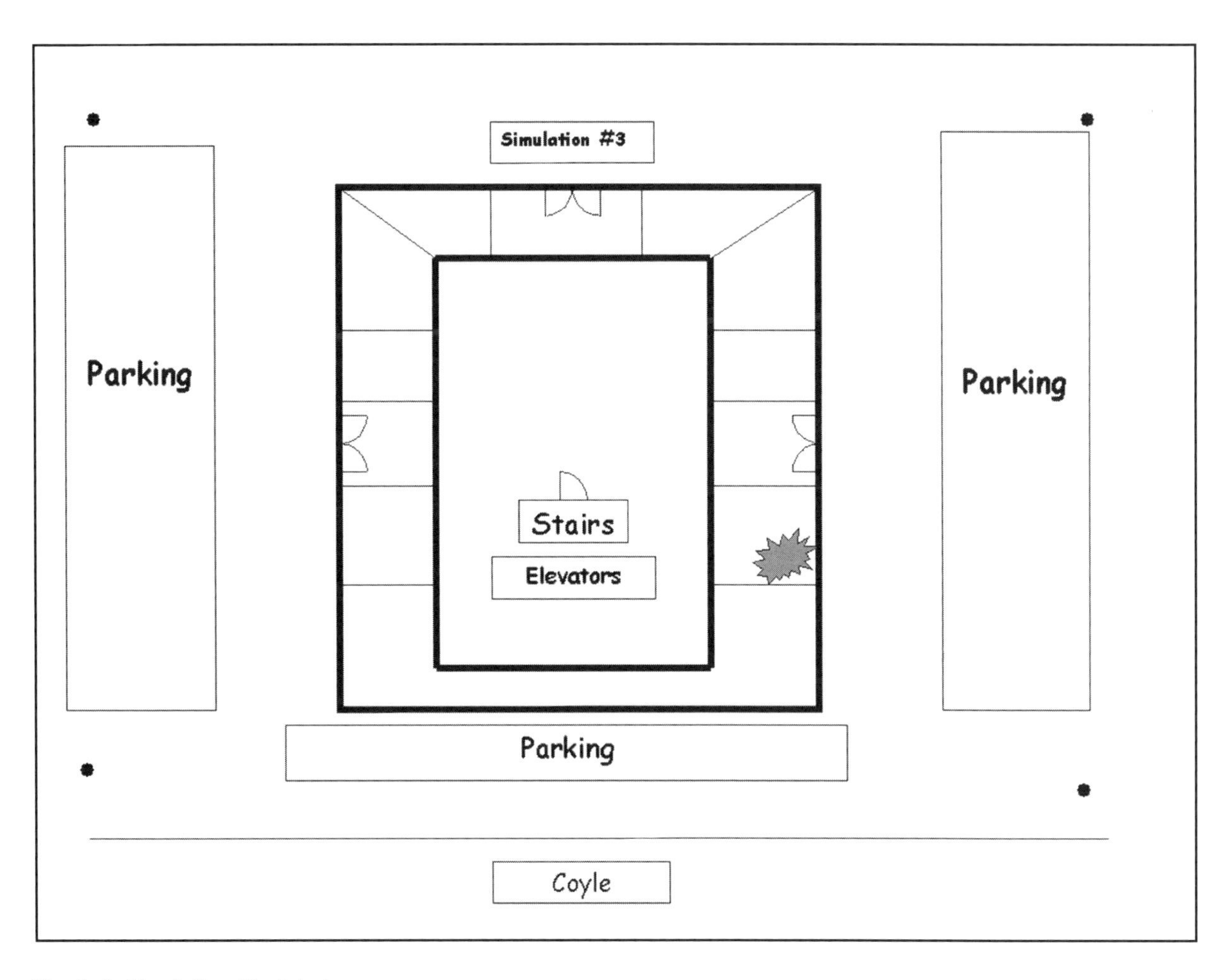

Fig. 6–6 Simulation #3 plot plan.

Emergency Simulation #3 Score Sheet

Candidate Name: ______________________________

Directions:

For each of the 10 dimensions below, score the candidate (1–10, with 1 being lowest, 10 being highest). Total score of 70 required for passing.

Problem-solving skills: **Score:** ______

(Defined current problems and could see early signs of pending problems; applied effective solutions in a safe and timely manner.)

Resource management skills: **Score:** ______

(Managed personnel, equipment, apparatus; called appropriate additional resources in a timely fashion; maintained accountability; gained most out of limited resources.)

Prioritization skills: **Score:** ______

(Took action and utilized resources based on priorities of (1) life safety, (2) loss control, and (3) customer service.)

Oral communication: **Score:** ______

(Effectively verbalized plan of action, thought process, and contingencies; utilized additional channels when appropriate.)

Composure: **Score:** ______

(Remained calm under pressure; maintained control of emotions regardless of situation.)

Strategy: **Score:** ______

(Articulated a strategy for solving the problem; verbalized offensive, marginal, and defensive strategies based on situation.)

Tactical skills: **Score:** ______

(Employed safe, effective, and efficient tactics based on clearly stated tactical objectives; utilized industry standard recognized procedures properly.)

Safety: **Score: ______**

(Maintained high priority toward safety; RIC established; did not put personnel in unsafe situations; had proper resources to support tactics.)

ICS: **Score: ______**

(Utilized the ICS components properly; maintained proper span of control; built system appropriate for the plan and the problem.)

Command presence: **Score: ______**

(Displayed confidence and leadership during the evolution that would inspire action and confidence of subordinates on the scene.)

Total Score: ______

Comments:

____

____

____

____

Emergency simulation exercise #4

In this exercise, you will be evaluated on your emergency scene management skills. During your assessment, you will be dispatched to an incident. You will have two minutes from dispatch before you arrive on scene. Upon arrival, a picture of the scene and a plot plan of the area will be placed before you on the screens.

You will be scored on the following dimensions:

- Problem-solving skills
- Resource management skills
- Prioritization skills
- Oral communication
- Composure
- Strategy
- Tactical skills
- Safety
- ICS
- Command presence

After arrival, you will have up to 10 minutes to manage the scene. After 10 minutes, you will perform a face-to-face pass on to the division chief portrayed by an assessor. You will have no longer than two minutes to perform the pass on. After completion of the simulation, the assessors will ask you questions regarding your performance.

You will have a pad of paper and pencils during your simulation to use as you see fit. In addition, you will have colored markers to utilize the plot plan and mark the placement of any apparatus and equipment during your simulation. Do you have any questions?

You must ask for additional resources via radio. All communications with other companies and dispatch will be via radio; however, you must verbalize your thoughts and actions when appropriate.

The time is 2 p.m. on Saturday. You are dispatched to a commercial structure fire at 8900 Greenback Lane. You are the Battalion 1. The additional units dispatched to this incident are Engine 1, Engine 2, Engine 3, Truck 1, and Truck 2. You will assume command upon arrival. Are you ready to begin? You have two minutes before you arrive.

(Fig. 6–7 shows the fire scene simulation window. Fig. 6–8 is the corresponding plot plan.)

Fig. 6–7 Simulation exercise #4.

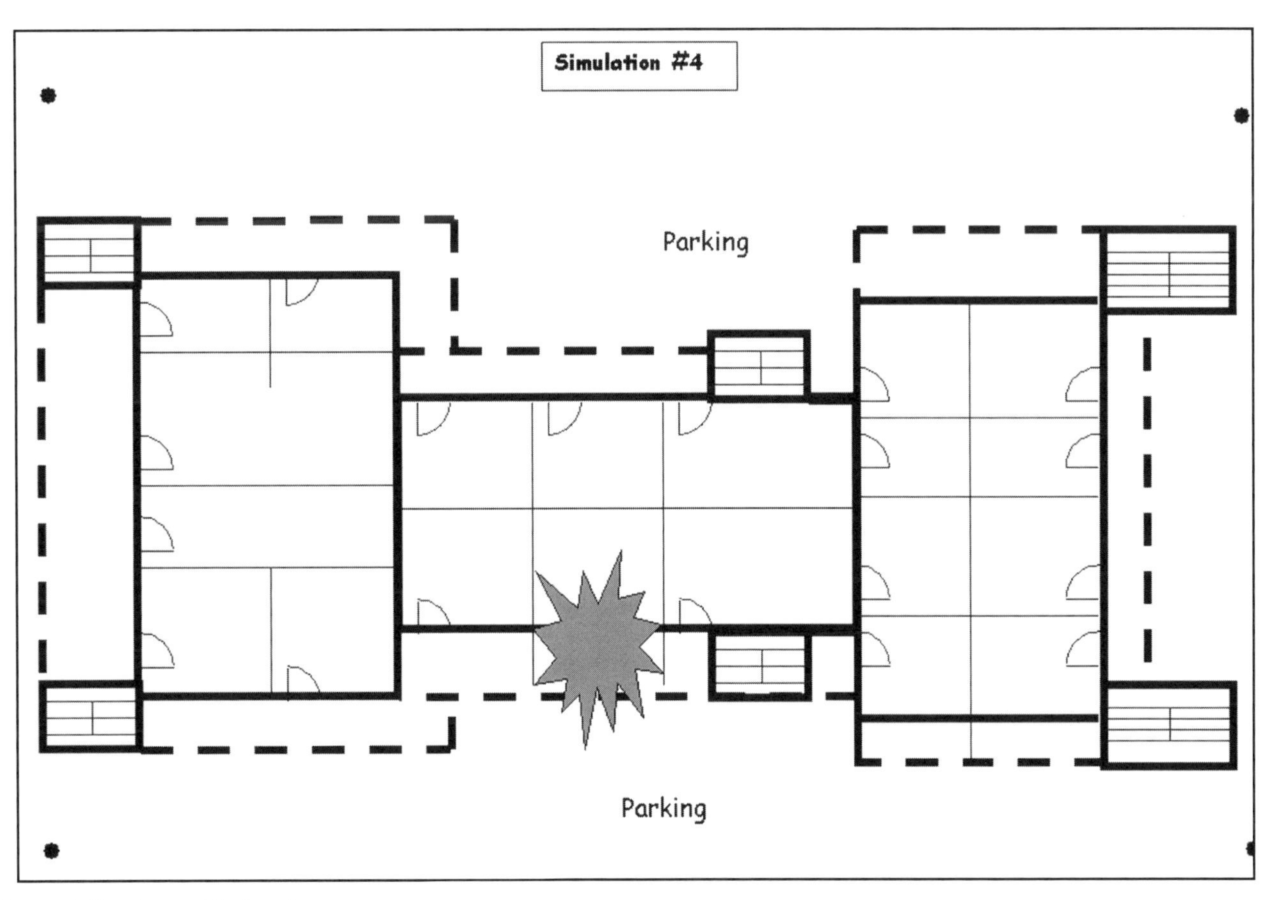

Fig. 6–8 Simulation #4 plot plan.

Emergency Simulation #4 Score Sheet

Candidate Name: ______________________________

Directions:

For each of the 10 dimensions below, score the candidate (1–10, with 1 being lowest, 10 being highest). Total score of 70 required for passing.

Problem-solving skills: **Score:** ______

(Defined current problems and could see early signs of pending problems; applied effective solutions in a safe and timely manner.)

Resource management skills: **Score:** ______

(Managed personnel, equipment, apparatus; called appropriate additional resources in a timely fashion; maintained accountability; gained most out of limited resources.)

Prioritization skills: **Score:** ______

(Took action and utilized resources based on priorities of (1) life safety, (2) loss control, and (3) customer service.)

Oral communication: **Score:** ______

(Effectively verbalized plan of action, thought process, and contingencies; utilized additional channels when appropriate.)

Composure: **Score:** ______

(Remained calm under pressure; maintained control of emotions regardless of situation.)

Strategy: **Score:** ______

(Articulated a strategy for solving the problem; verbalized offensive, marginal, and defensive strategies based on situation.)

Tactical skills: **Score:** ______

(Employed safe, effective, and efficient tactics based on clearly stated tactical objectives; utilized industry standard recognized procedures properly.)

Safety: **Score: ______**

(Maintained high priority toward safety; RIC established; did not put personnel in unsafe situations; had proper resources to support tactics.)

ICS: **Score: ______**

(Utilized the ICS components properly; maintained proper span of control; built system appropriate for the plan and the problem.)

Command presence: **Score: ______**

(Displayed confidence and leadership during the evolution that would inspire action and confidence of subordinates on the scene.)

Total Score: ______

Comments:

____

____

____

____

In-Basket Exercise

You are a captain on Engine 1. Today is August 2. The time is 8:15. You just began your first shift back after four days off. After this shift, you will be on vacation for the next two shifts

You have the combined e-mails, memos, and phone messages as follows waiting for you. Review all of the items and take appropriate actions. After processing all of the items below, you will present your response to each issue to the assessment panel. If your response includes a written correspondence, you will be required to actually write the memos/e-mails that you would utilize in dealing with these issues.

You will give a presentation of your actions, and any written correspondence to the panel in two hours. For each item, be prepared to describe the action you would take and who you would contact and give any written documentation to the raters. You will have 15 minutes to give your presentation. Questions may be asked during your presentation.

Phone messages

Figures 6–9 through 6–20 are the phone messages waiting for this exercise. The exercise e-mails follow.

NAME: *BC Cranky*
DATE: *8/2/04*
TIME: *08:00*
MESSAGE: *The Battalion Chief from the off-going shift wants to discuss a call from yesterday. He may need you to go by the house of an RP to get more info regarding a fire.*

Fig. 6–9 This first message could be a higher priority due to the time element of 0800, if that is your shift change time. Remember, do not act on any one message until you prioritize the entire stack.

NAME: *Captain Crazy*
DATE: *8/2/04*
TIME: *07:52*
MESSAGE: *E2 wants to know if you want to have a wet drill this afternoon. It's a required drill this month.*

Fig. 6–10 This should be a lower priority since you do not have to respond until this afternoon.

NAME: *AC Robinson*
DATE: *8/1/04*
TIME: *14:30*
MESSAGE: *Please call the training division to set up the academy drill.*

Fig. 6–11 Since this is already a day old, it should be a lower priority.

NAME: *Joe Phillips*
DATE: *8/1/04*
TIME: *5:30pm*
MESSAGE: *EMT ride-along tomorrow. Should be here at 2pm.*

Fig. 6–12 This is a lower priority due to the afternoon time frame. You should delegate the responsibility of supervising the ride-along to one of your firefighters.

NAME: *Alice Davies*
DATE: *7/30/04*
TIME: *10:00*
MESSAGE: *Look forward to the station tour for the kids on the 2nd. Please call if there is a change.*

555-1212

Fig. 6–13 The tour is today, but Mrs. Davies did not indicate when the tour is taking place. This could be a high priority. You must call to find out the time as it could be first thing in the morning.

NAME: *Fleet division*
DATE: *7/30/04*
TIME: *11:15*
MESSAGE: *Engine 1 is due for servicing on the 4th. Please have at shop by 09:00.*

Fig. 6–14 This is a lower priority since it isn't due to the shop until next shift. You must ensure the message is passed on to whoever is working for you that day since you are on vacation.

NAME: *Fire Chief*
DATE: *7/29/04*
TIME: *08:10*
MESSAGE:
Sorry he missed you this morning. Please call his office as soon as you get back from your 4-day.

Fig. 6–15 Now that you are back from your 4-day, you should call him. Since this is the fire chief, it's a high priority.

NAME: *Thomas Raley*
DATE: *7/29/04*
TIME: *10:15*
MESSAGE: *Look forward to the fire re-inspection. Should have everything done for appointment on the 2nd at 9am.*

Fig. 6–16 You now have a fire inspection at 9 a.m. This is a high priority and could potentially conflict with other issues this morning.

NAME: *Bob Taylor*
DATE: *7/29/04*
TIME: *09:20*
MESSAGE: *Wants to know if you still need help with recruit training.*

Fig. 6–17 This is a low priority since the date is 7/29 (4 days old). It is related to the message from AC Robinson about the academy. Keep this in mind or clip the items together.

NAME: *Fire Chief*
DATE: *7/30/04*
TIME: *12:15*
MESSAGE: *Will be by to see you and the crew at 9am on the 2nd.*

Fig. 6–18 This is related to memo in figure 6–15. Now you know that he is coming by at 9 a.m. today. This could conflict with the tour (see fig. 6–13) and definitely conflicts with the inspection (see fig. 6–16).

NAME: *Florence Taylor*
DATE: *7/30/04*
TIME: *9pm*
MESSAGE: *Please call about Bob ASAP.*

Fig. 6–19 This is about Bob Taylor (fig. 6–17). You should deduce this from her last name. It sounds possibly urgent.

NAME: *Thomas Raley*
DATE: *7/30/04*
TIME: *11:00am*
MESSAGE: *Unable to get extinguishers serviced by the 2nd. Please call to reschedule a re-inspection for the 4th.*

Fig. 6–20 This memo is a relief since it eliminates the conflict from memo in figure 6–16. Now, you are available to meet the chief at 9 a.m. As you look at the e-mails, process them the same way. Prioritize and look for related/conflicting items.

E-mails

August 2, 07:30
Good morning Captains. Welcome back from a great 4-day. Please make sure that you complete all of your incident reports before going home at the end of shift.

Today's training is up to you. We have a multi-company wet drill required this month. If you have the drill today, please let me know so that I can evaluate.

Other than that, be careful out there!

BC Cranky

August 2, 07:15
Hey Captain, I will be late this morning. I am working an overtime shift at Station 53 across town. I should be there by 8:30. I called but A shift was out on a run, so I couldn't leave a message.

FF Ransom

August 1, 11:07
To Captains on C and B shift, E1 is still having overheating problems. We tried to change out to a reserve engine yesterday but had no luck due to high call volume. We also had a structure fire tonight but the engine seemed ok. I don't know if we would have had problems if it were in the middle of the day. Keep an eye on it.

A Shift Captain – E1

August 1, 14:35
Please call the training division to set up a forcible entry class for the academy on August 10. Bob Taylor will be unable to assist. He will be out of town due to family issues. You are responsible for setting up instructors, providing curriculum, and setting up an exam. Thank you for your help. The last class was great!

AC Robinson

August 1, 10:30
Captains, please notify your personnel that overtime slips must be filled out for anyone who holds over for more than 15 minutes. The slips must be electronically filed prior to leaving the shift. Also, have your vacation slips in at least 24 hours prior to taking your vacation. We have had some problems with people not filling out paperwork until the last minute.

Thank you,
Finance Division

July 30, 10:00
Hey Captain, just sent you a phone message about the 2nd. The kids are so excited to come to the station! We plan to be there at about 9:30. Let me know if you need any more information.

Alice Davies, 555-1212

July 30, 09:53
Are you interested in the Sacramento Bee? Just Reply to this e-mail or call **1-800-BEE-READ!**

July 30, 09:45
All companies, your Fire Prevention Bureau will be celebrating Fire Prevention Week August 7–13! Please hand out stickers and coloring books whenever possible.

Thank you
FPB

July 29, 09:19
Just a friendly reminder that station safety inspections are due by August 2. Please complete the attached form and send back in the department mail.

Thank you. Safety Division

July 29, 08:17
The union meeting on August 3 has been cancelled. We will reschedule. Date TBA

Keep the faith!
Union Rep. Emmitsburg

July 29, 07:45
To all Company Officers –

The fire chief will be visiting all the crews as part of the annual "finger in the wind" sessions. Either I or the chief will be calling ahead to arrange an appointment. Thanks in advance for your time!

Marilyn Reality, Office of the Fire Chief

July 29, 02:00
Captain, I was hoping to finish our talk today, but we were both too busy. Please give me a call next shift so I can tell you what's going on. I do not want this to be a big deal. I just want it to stop. I'll be at Engine 2. Thank you.

FF Sylvia Mayorga

July 28, 23:46
The annual summer association picnic is coming up. Please buy your tickets soon as they are going fast! Remember, the date is August 13 at Grace Davis Park. Call Terry at E3B for information on how you can help and purchase tickets. We need help with set up, BBQ, raffle, and the softball game. Anyone with extra gloves please bring them. This will be fun for the whole family!

Terry Gianni – E3B

Modified In-Basket Exercise

You are the newly promoted captain on Engine 4, C shift. Today is the 15th of the month, and the first shift back from a four-day. Your station has an engine, medic unit, and a grass unit. You have a crew of five (you, an engineer of 20 years, a firefighter of 10 years, and two probationary firefighter/paramedics on the medic unit).

Review and process the events listed below. After one hour, you will meet with the assessment panel. You will have 20 minutes to present your actions and correspondence (including who you will contact) for all of the events. For each event, state what you would do, whom you would contact (if anyone), and how you would communicate with them. You will then be asked questions for up to 10 minutes after your presentation.

- 07:45 – Engineer calls in sick for the third time this month.
- 08:03 – Off-going firefighter wants to talk to you about her captain on B shift.
- 08:19 – Emergency call — medical aid.
- 08:52 – Return to station. Callback engineer is at station.
- 09:15 – Callback engineer states that rear flashing lights do not work on the engine.
- 10:10 – BC calls and wants to conduct a drill at 13:00.
- 10:30 – E-mail from Mrs. Smith from Silly Elementary School. She wants to know if you will bring the engine by at 13:30 since B shift blew her off twice. B shift told her that you were the one to contact for these events.
- 10:35 – Emergency call — vehicle accident.
- 11:30 – Probationary firefighter burns himself with scalding hot water while making lunch for the crew.
- 14:30 – EMT ride-along arrives. She has never been in station before.
- 14:45 – Captain from E3 calls, asks if you would conduct a drill regarding the new gurneys like the one on your medic. He's worried about back injuries.
- 15:00 – Firefighter Sylvia Mayorga calls to finish conversation started about her captain. She feels uncomfortable at the station and feels he is unfair and will not do anything about the others on the crew who are making her feel harassed. She does not feel comfortable talking to him about it.
- 15:20 – Emergency call — miscarriage.
- 16:37 – BC wants to come by to talk about vehicle accident. He felt some things could have been better. He wants to discuss as a crew. He will come by at 19:00.
- 17:30 – During dinner, doorbell rings. A neighbor down the street wants a blood-pressure check.
- 18:45 – Emergency call — grass fire in vacant field.
- 20:30 – BC stops by after grass fire. He wants to discuss the vehicle accident from earlier that shift.
- 23:00 – Ride-along still in quarters.
- 02:30 – Emergency call — teenage hanging.
- 07:30 – B shift captain arrives to work.

In-Basket/Modified In-Basket Exercise — Score Sheet

Candidate Name: ______________________________

Score: 1 – Performs task with poor ability
2 – Performs task with below-average ability
3 – Performs task with average ability
4 – Performs task with above-average ability
5 – Performs task with outstanding ability

Prioritization skills:

- Maintained priorities of (1) safety, (2) effectiveness and (3) efficiency – 1 2 3 4 5
- Placed high priority on personnel issues – 1 2 3 4 5
- Not distracted by trivial issues – 1 2 3 4 5
- Able to recognized urgent vs. non-urgent issues – 1 2 3 4 5

Delegation skills:

- Utilized crewmembers to assist with tasks – 1 2 3 4 5
- Saw teaching and learning opportunities – 1 2 3 4 5
- Did not over-delegate tasks – 1 2 3 4 5
- Did not under-delegate tasks – 1 2 3 4 5

Problem-solving skills:

- Identified related issues – 1 2 3 4 5
- Followed relevant policy – 1 2 3 4 5
- Able to identify conflicting issues early – 1 2 3 4 5
- Attempted to solve multiple problems with one solution – 1 2 3 4 5

Time management:

- Performs tasks within allotted time frame – 1 2 3 4 5
- Does not seem rushed – 1 2 3 4 5
- Allows time to answer questions – 1 2 3 4 5

Planning skills:

- Effectively planned day's events – 1 2 3 4 5
- Seemed well organized, followed a logical sequence – 1 2 3 4 5
- Accomplished tasks and goals – 1 2 3 4 5

Leadership:

- Shows strong initiative – 1 2 3 4 5
- Took action where appropriate – 1 2 3 4 5

Total Score:________

Comments:

____

____

Role-Play/Counseling Exercises for Captains and Lieutenants

Role-play/counseling exercise #1

Directions: The role-play/counseling exercise will evaluate your interpersonal, leadership, oral communications, time management, and problem-solving skills. You will participate in an exercise with an individual who portrays a member of your crew. You are expected to participate fully and be prepared to answer assessor questions about the session.

You will have five minutes to review the situation described here. You then will have 10 minutes to discuss the issues with your crewmember. Afterward, the assessors will ask questions for up to five minutes regarding your performance.

Situation: You are a newly promoted captain at Engine 1. You have been with the department for 10 years. Your engineer, Dan, has been with the department for 20 years.

Ever since you were assigned to the engine, Dan seems to go out of his way to criticize you. You can't seem to do anything right in his eyes.

He rides you when you go on calls. When you try to give him directions from the map book, he tells you that he knows where he's going, and that you don't need to tell him how to drive. More than once, he has gotten lost after he said, "I know where I'm going. You just sit there and look important."

He gets along well with everyone else, especially his former captain, who just retired. Everyone thinks he's great to work with and very knowledgeable. You just seem to be the exception.

Other examples of his apparent disrespect include interrupting you during drills by telling the probationary firefighter how it works better his way; leaving to make personal business calls during morning meetings; and falling asleep during training videos.

You have asked him to sit down and have a talk. It's 9 a.m. at Station 1.

Role player #1 directions: You will participate in a role-play/counseling exercise. Please make every attempt to maintain the character described as follows. Do not volunteer any information unless you are led to or directly asked to. Make the candidate work to get information out of you, but let it out when appropriate.

Your name is Dan. You are an engineer on Engine 1. You have been with the fire department for 20 years, seven as an engineer.

You enjoy the job very much and have attempted to promote to captain for 10 years. Although you have great natural leadership abilities, you can't seem to pass the assessment center. This has caused you great frustration, especially since your new captain has been with the department only 10 years. You remember when he was hired. That was when you took your first captain test. Now, five tests later, you are still an engineer and he made captain on his first attempt. He is very energetic and seems to know his stuff, but you have a hard time adjusting to working for someone younger who got the job without having to pay his dues.

Your former captain retired. You and he were together since you were promoted to engineer. He taught you a lot about the job, but he couldn't seem to help with the testing processes they use today. Now you feel like you will never get promoted, and the young kids are passing you by. In fact, you spend more time worrying about your construction business on the side. The only satisfaction around the firehouse is helping to teach the probies. At least they could learn something from you. You will meet with your captain for 10 minutes.

Role-play/counseling exercise #2

Directions: You will participate in a counseling exercise that involves a role player. The role-play/counseling exercise will evaluate your interpersonal, leadership, oral communications, time management, and problem-solving skills.

You will have five minutes to review the situation here. You then will meet with your subordinate for 10 minutes. During that time, the assessors will evaluate your performance. Finally, you will be given five minutes to answer questions regarding your performance.

Situation: You are a captain on Engine 3, A shift. You have an engine and medic unit assigned to your house. In addition to you, the engineer, and firefighter on your engine, your medic unit is staffed with two probationary firefighters.

Your engineer, Sylvia Mayorga, was promoted a month ago and was assigned to your engine at that time. She came to you last shift and stated that she felt uncomfortable with the colorful language of your senior firefighter, George Reed. He always seems to notice the female attributes of your patients, citizens at the store, and women on TV.

Engineer Mayorga does not want to make big waves since she's new to the crew, but she definitely wants Firefighter Reed's comments to stop. She feels they are inappropriate. She does not feel comfortable talking to him.

Firefighter Reed is a good friend. The two of you were in the academy together 13 years ago. He has no aspirations for promotion but is a solid firefighter. He is great on drills, loves to cook for the crew, and always is the first one to step up when a special project needs to be done.

After talking to Engineer Mayorga, you have decided to have a meeting with Firefighter Reed.

Role player #2 directions: You will participate in a role-play/counseling exercise. Please make every attempt to maintain the character described as follows. Do not volunteer any information unless you are led to or directly asked to.

Your name is George Reed. You are a firefighter on Engine 3, A shift. You have been a firefighter with your department for 13 years.

You enjoy work, love to help out, and get a lot out of teaching the two new probies at your station. You have no real aspirations for promotion because you enjoy your job and love your captain. Since the two of you were in the academy together, you have a special bond. You would love to spend the rest of your career with your captain.

Your home life isn't so great; however. Your wife of eight years has left, and you are sure this will lead to divorce. Fortunately, you have no kids although you want some one day. You are trying to keep a good attitude at work despite your personal life. You can't wait for the divorce to be final so you can start dating again. No one knows about your divorce at work. Not even the captain.

Once again, do not volunteer any information. The captain must ask questions that will lead him/her to the information described above. Only give information when asked or you feel that you are led to elaborate on the situation.

Role-play/counseling exercise #3

Welcome to the subordinate counseling exercise. You will be taking part in an interactive role-play exercise. The areas evaluated will be:

- Oral communications
- Problem solving
- Time management
- Interpersonal skills
- Leadership

You are the captain on Truck 4. You have two seasoned firefighters on your company and a great salty engineer. Last shift, you responded to a structure fire and performed vertical ventilation, among other tasks. You were on the roof with one of the firefighters, while the engineer and other firefighter performed ground-level operations.

The fire was confined to the attic, as it started from an electrical problem in the attic wiring. As far as you were concerned, the truck operations went very well. In fact, you took the company out for ice cream sundaes after dinner that night.

This morning, the BC called. He states that the homeowner of that house, Mr. Charles Rumson, was furious about the broken windows around the house. He called the fire chief and chewed him out for the "reckless tactics and unnecessary damage" to his home. He was not satisfied with talking to the chief. He wanted to talk to the "captain in charge of the crew who did such sloppy work." The BC states that in fact he saw your firefighter take the windows out, but he didn't think much of it at the time since he was focused on the fire. The BC will be in an operations meeting today. He made an appointment for Mr. Rumson to see you today at 1 p.m.

Role player #3: Your name is Charles Rumson. You had a structure fire in the attic of your home 2 days ago. You were happy that the fire department did such a good job on confining the fire to the attic, but you are *absolutely irate* at the fact that they broke almost all of the windows in your home. You did not understand why they did that since the fire was confined to the attic space.

Before you retired from the fire service, you were a truck captain for 10 years in Los Angeles (LA) County. You prided yourself in never doing any more damage than necessary to a building. Your motto was, "Treat every home like it's your home."

You called the fire department and spoke to the chief. You insisted on talking to the captain on the truck to teach him a lesson or two about damage to property. The chief had the BC call you and arrange a meeting with the captain on Truck 4. You are absolutely disgusted and can't wait to lay into the truck captain. Be convincing in your role. You are attempting to get an admission of poor judgment. You do not want anything more than to know that the captain won't let it happen again to someone else's home. You share your philosophy with him.

Role-play/counseling exercise #4

Welcome to the subordinate counseling exercise. You will be taking part in an interactive role-play exercise. The areas evaluated will be:

- Oral communications
- Problem solving
- Time management
- Interpersonal skills
- Leadership

You are the captain on Truck 4. You have a great crew that you trust: two seasoned firefighters and a salty engineer.

Last shift, you had an attic fire in which you performed ventilation and other typical truck functions with your crew. You thought everything went well until you had a meeting with the homeowner (Mr. Rumson) today. He was absolutely irate about the fact that your crew broke nearly all of the windows in his home when the fire was only in the attic area. He was a truck captain for 10 years in LA County and gave you an earful about the "right" way to do things, especially salvage.

You were on the roof when the firefighter took out the windows. You assured Mr. Rumson that you would make sure it never happened again and that you will talk to the firefighter who did this.

Your firefighter, Rudy Montoya is an excellent up-and-comer type of firefighter. He is #1 on the captain's list and will make a good captain some day. He needs a little refining, as he can be a bit aggressive on calls. This is a good example of you having to clean up after Rudy sometimes. In fact, other examples include his jumping the chain of command from time to time and taking over on drills with other companies. You know he means well, but you are trying to help him be a bit wiser.

You have called a meeting with Firefighter Montoya to discuss the attic fire and your discussion with Mr. Rumson. You will meet with Firefighter Montoya in 5 minutes. You will then have 10 minutes to discuss the situation with Firefighter Montoya. After that time, the assessors will ask you questions regarding your actions.

Role player #4: Your name is Firefighter Rudy Montoya. You have been with the fire department for 11 years. You are currently assigned to Truck 4. You are currently #1 on the captain list for promotion.

Last shift, your company responded to an attic fire. While the fire was being knocked down in the attic, you found a good opportunity to teach one of the probationary firefighters on the engine how to properly break windows. You are passionate about training and love to think of new and exciting ways to teach.

The probationary firefighter looks up to you, and you enjoy teaching him at every opportunity. Since you hope to be a captain soon, you love mentoring and can't wait to have your own crew some day.

You are on the urban search and rescue (USAR) team, teach Rescue Systems 1 and 2, and are working on the new water rescue program for your department.

The captain has called you in to discuss the fire last shift. You are sure that he wants to give you extra gold stars in private for utilizing the fire as a training opportunity.

You are glad that he is your captain since he is one of the few people who "get" you. Everyone else just seems a little too sensitive or just threatened by your attitude.

Be sure to maintain your character. You are a high-energy person who likes to talk but has a hard time hearing what people are trying to tell you sometimes. You have a good heart, but you need a little mellowing.

Role-Play/Counseling Exercises for Chief Officers

Role-play/counseling exercise #5

You are the new BC of Battalion 12. You have been in your new position for three months. You have noticed that on all fires, Captain Ivan Reckless does not wear his breathing apparatus or helmet. In fact, you noticed that his probationary firefighter is starting to follow his poor example.

You have counseled him before on the fire scene (in private) when the fire was in the overhaul phase. He repeatedly has told you that he has been in the fire service longer than you and plans to retire in three months. He feels entitled to not wear his protective clothing if he chooses. After all, he made it almost 30 years without your help.

After the latest house fire, you noticed that he is still not wearing his gear. You have decided to meet with him to discuss it. You will meet in 10 minutes, and then have 10 minutes to discuss the issue.

Role player #5: Your name is Captain Ivan Reckless. You are three months away from retiring from a 30-year career. You have a probationary firefighter in your crew and are inconvenienced by his presence.

You cannot wait to retire. You are just trying to get to retirement so you can relax. You look forward to fishing and being with your wife and grand kids. You are just tired of the job.

Role-play/counseling exercise #6

You are the new BC at Battalion 3. You have had numerous complaints about engine companies in your battalion that are driving recklessly. The engines are running red lights when not responding to calls and have yelled at citizens for not getting out of the way sooner. In fact, one officer, Captain Short Fuse pulled over a driver when responding to a call and publicly berated her on the side of the road. You already counseled Captain Fuse and he apologized for the event.

The complaints were given to you from the division chief before you held your first officer meeting. You have called your first captain meeting with your new officers to share your expectations and discuss these complaints. You will meet with them in 10 minutes and have 10 minutes to discuss your expectations and the issue described.

Role Play/Counseling Exercise — Score Sheet

Candidate Name: ______________________________

Score: 1 – Performs task with poor ability
2 – Performs task with below-average ability
3 – Performs task with average ability
4 – Performs task with above-average ability
5 – Performs task with outstanding ability

Interpersonal skills:

- Maintains calm and professional attitude – 1 2 3 4 5
- Creates safe environment – 1 2 3 4 5
- Patient with individual(s) – 1 2 3 4 5
- Firm, fair, and friendly – 1 2 3 4 5
- Respectful – 1 2 3 4 5
- Seeks to resolve conflict – 1 2 3 4 5
- Positive attitude and solution oriented – 1 2 3 4 5
- Separates person from behavior – 1 2 3 4 5

Oral communication skills:

- Speaks clearly and effectively – 1 2 3 4 5
- Listens attentively, attempts to hear message – 1 2 3 4 5
- Conveys message in terms listener understands – 1 2 3 4 5

Problem-solving skills:

- Able to see through symptoms to define problem – 1 2 3 4 5
- Defines problem accurately – 1 2 3 4 5
- Considers alternate solutions – 1 2 3 4 5
- Able to implement solution for effective results – 1 2 3 4 5
- Articulates monitoring systems for improvement – 1 2 3 4 5

Leadership:

- Able to motivate individual to find solution – 1 2 3 4 5
- Provides direction and remains on task – 1 2 3 4 5

Time management:

- Performs tasks within allotted time frame – 1 2 3 4 5
- Allows time to listen as well as resolve situation – 1 2 3 4 5

Total Score: ________

Comments:

____

____

Oral Presentations

Oral presentation #1

You are a captain on Engine 5. You have decided to conduct an impromptu drill with your crew regarding fire station safety. You have 10 minutes to prepare a 10-minute presentation. Your crew will then have five minutes to ask questions.

KSAs evaluated:

- Oral communication skills
- Interpersonal skills
- Team-building skills
- Time management skills
- Planning skills
- Leadership skills

Oral presentation #2

You are a BC for Dream Fire Department. The fire chief has asked you to give a presentation to a community group regarding the department's preparation for weapons of mass destruction and terrorism readiness. You have 20 minutes to prepare a 10-minute presentation. The community group will then have five minutes to ask you questions.

Oral presentation #3

You are the captain on Engine 10. You have decided to conduct an impromptu drill with your crew about the importance of customer service. You have 10 minutes to prepare a 10-minute presentation. They will ask you questions at the conclusion.

Oral presentation #4

As a BC, you have 90 minutes to write a five-page paper on a critical issue facing your department. Then spend 20 minutes preparing a 10-minute presentation for your fire chief.

Oral presentation #5

You are asked to deliver an oral/visual resume. You have 20 minutes, an easel pad and colored markers to make your presentation. You will have no longer than five minutes to give the presentation.

Oral Presentation Exercise — Score Sheet

Candidate Name: ______________________________

Score: 1 – Performs task with poor ability
2 – Performs task with below-average ability
3 – Performs task with average ability
4 – Performs task with above-average ability
5 – Performs task with outstanding ability

Oral communication skills:

- Speaks clearly and effectively – 1 2 3 4 5
- Conveys message in terms the listeners understand – 1 2 3 4 5
- Includes all listeners in the presentation – 1 2 3 4 5
- Gains listener interest early – 1 2 3 4 5
- Responsive to feedback – 1 2 3 4 5

Interpersonal skills:

- Maintains calm and professional attitude – 1 2 3 4 5
- Respectful – 1 2 3 4 5
- Seeks to educate – 1 2 3 4 5
- Positive attitude – 1 2 3 4 5

Team-building skills:

- Brought group together toward common goal – 1 2 3 4 5
- Promoted discussion where appropriate – 1 2 3 4 5
- Found common interests – 1 2 3 4 5

Time management:

- Performs tasks within allotted time frame – 1 2 3 4 5
- Does not seem rushed – 1 2 3 4 5
- Allows time to answer questions – 1 2 3 4 5

Planning skills:

- Effectively planned presentation – 1 2 3 4 5
- Presentation seemed well organized, followed a logical sequence – 1 2 3 4 5

Leadership:

- Shows strong initiative – 1 2 3 4 5
- Inspires listeners to take part in communication – 1 2 3 4 5
- Appears confident in front of the group – 1 2 3 4 5

Total Score:________

Comments:

____

____

Supervisory Scenarios for Captains and Lieutenants

You have 15 minutes to review the following four issues. You will then give a 10-minute presentation to the assessors regarding the actions you would take. You will be sitting at the table across from the assessors. No visual aids will be used.

Scenario #1

You are on the scene of a fire. Your firefighter yells at a bystander about getting in the way. You hear the firefighter call the bystander an "idiot," and then the firefighter yells at him for being "stupid" while stretching hose. The bystander is visibly upset.

Scenario #2

While at the grocery store, a woman openly criticizes you and your crew for "wasting taxpayer dollars" while shopping.

Scenario #3

A member of your crew is having a hard time with the firefighter on B shift. He is constantly complaining about the engine being in disarray at shift change. You notice that sometimes the engine is not properly placed back in service, as he describes. Other times, the engine is fine but your firefighter insists that it's getting worse each cycle.

Other issues include your firefighter openly leaving pictures of the B shift firefighter around the station with disparaging names and knives going through his head.

Scenario #4

Your BC sends you an e-mail stating that your engine ran a red light while you were on vacation last week. He wants you to handle.

Supervisory Scenarios for Battalion Chiefs

It's your first day on the job in your new position as BC. You get the following e-mails. Review all of them within 20 minutes and then be prepared to discuss your actions with the assessors. You will have 10 minutes to present your actions.

Scenario #1

Hey Chief, we need to get together. My probationary firefighter is not going to make it. She is having problems throwing the 24-foot ladder and anything involving upper body strength is almost impossible. I recommend that she get transferred to another battalion or engine that is not as busy. I just do not have time for this.

Capt. Billy Bob, Engine 12

Scenario #2

Welcome to the battalion boss, I hope you enjoy it here. I will be sending you a copy of the counseling memo I wrote for Sam Simpson. He is late to work almost every other shift. I recommend that we take this to the next level. Please call me.

Captain Joey Jones, Engine 13

Scenario #3

Chief, please look into a citizen complaint about excess damage at a house fire. The citizen's name is Freddy Filo. The crew in question was Captain Billy Bob. Let me know your findings.

Assistant Chief Gomez

Scenario #4

Chief, just thought you should know that this firehouse is out of control. I am tired of the verbal abuse and constant griping by everyone, especially Captain Jones. He doesn't seem to care or do anything about the negativity. Since we have a probationary firefighter here, I thought you should know.

Specifically, there was an incident last week when the engineer yelled at the probie about loading the hose wrong. Captain Jones just laughed it off. I don't think that's right. We haven't had a regular BC here for a year, so I hope you can do something. Thank you.

Sam Simpson

Scenario #5

Hey, welcome to Battalion 3! We have a good time here. Just to let you know, I've been getting complaints about engine 12 leaving their rig a mess on a regular basis. We haven't had a regular BC here to follow up. I will see you next week when I get back from fishing. Good luck.

BC Green

Scenario #6

Please follow up on a citizen complaint that there was improper language and pictures in the fire station during a station tour. The mother of the scout troop stated that the firefighters didn't seem to want her and the scouts at the station. They were watching a sexually graphic movie during the tour, and some of the other mothers were very upset. Please investigate and write a letter to her for me to review. This was Engine 12's crew.

Assistant Chief Gomez

Supervisory Exercise — Score Sheet

Candidate Name: ______________________________

Score: 1 – Performs task with poor ability
2 – Performs task with below-average ability
3 – Performs task with average ability
4 – Performs task with above-average ability
5 – Performs task with outstanding ability

Initiative:

- Willing to take appropriate action – 1 2 3 4 5
- Action taken in timely fashion – 1 2 3 4 5
- Willing to make tough decisions – 1 2 3 4 5

Oral communication skills:

- Speaks clearly and effectively – 1 2 3 4 5
- Speaks with confidence and assertiveness – 1 2 3 4 5

Problem-solving skills:

- Able to see through symptoms to define problem – 1 2 3 4 5
- Defines problem accurately – 1 2 3 4 5
- Considers alternate solutions – 1 2 3 4 5
- Able to implement solution for effective results – 1 2 3 4 5
- Connected related issues and personnel – 1 2 3 4 5
- Articulates monitoring systems for improvement – 1 2 3 4 5

Goal setting

- Set up plan – 1 2 3 4 5
- Had timeline for resolution – 1 2 3 4 5
- Realistic, measurable, attainable goals – 1 2 3 4 5

Leadership:

- Able to motivate individuals to find solution – 1 2 3 4 5
- Provides direction and remains on task – 1 2 3 4 5
- Positive attitude and solution oriented – 1 2 3 4 5
- Separate person from behavior – 1 2 3 4 5

Time management:

- Performs tasks within allotted time frame – 1 2 3 4 5
- Allows time for questioning – 1 2 3 4 5

Total Score: ________

Comments:

____

____

Leaderless Group Exercise for all Ranks

The fire department is about to establish a new set of SOGs to get everyone on the same page operationally. The fire chief hopes that this will enhance safety and efficiency and provide better risk management.

You have been chosen by the fire chief to be part of a group charged with establishing a list of no more than 10 of the most critical topics for SOGs facing your department. You will have 30 minutes to meet with the other members of the group to come up with 10 topics for SOG development. The specific SOGs you choose will be developed at a later time by a separate group(s).

Leaderless Group Exercise for Chief Officers

The fire chief has asked all of his chief officers to get together to perform a SWOT analysis of the department. He would like to know what the current strengths and weaknesses are internally; and what the opportunities and threats are externally.

You will have 60 minutes to meet with the other chief officers to discuss and formulate a final report for the fire chief to review. He would like no more than a two page summary.

For the purposes of the assessment, all candidates will write a two page summary of the session. You will have an additional 60 minutes to write your summary after the 60-minute group session.

Leaderless Group Exercise — Score Sheet

Candidate Name: ______________________________

Score: 1 – Performs task with poor ability
2 – Performs task with below-average ability
3 – Performs task with average ability
4 – Performs task with above-average ability
5 – Performs task with outstanding ability

Teamwork:

- Builds consensus – 1 2 3 4 5
- Values all participants – 1 2 3 4 5
- Ensures no one is left out of process – 1 2 3 4 5

Oral communication skills:

- Speaks clearly and effectively – 1 2 3 4 5
- Listens attentively, attempts to hear message – 1 2 3 4 5
- Conveys message in terms listener understands – 1 2 3 4 5

Problem-solving skills:

- Able to see through symptoms to define problem – 1 2 3 4 5
- Defines problem accurately – 1 2 3 4 5
- Considers alternate solutions – 1 2 3 4 5
- Able to implement solution for effective results – 1 2 3 4 5
- Articulates monitoring systems for improvement – 1 2 3 4 5

Goal setting

- Set up plan – 1 2 3 4 5
- Had timeline for resolution – 1 2 3 4 5
- Realistic, measurable, attainable goals – 1 2 3 4 5

Leadership:

- Able to motivate individuals to find solution – 1 2 3 4 5
- Provides direction and remains on task – 1 2 3 4 5
- Positive attitude and solution oriented – 1 2 3 4 5
- Attempts to keep all members engaged – 1 2 3 4 5

Interpersonal skills:

- Does not attempt to dominate situation – 1 2 3 4 5
- Demonstrates patience with differing opinions – 1 2 3 4 5

Total Score: ________

Comments:

____

____

Additional Key Points and Common Pitfalls 7

Time Management

Time management is one of the biggest areas candidates fail. The reality is that you must manage yourself. You must keep track of your time and discipline yourself during your assessment center.

Each exercise will have a time limit. In addition, you must keep track of time in between exercises. Many exercises will allow a preparation time as well. Keep track of the time at all times!

Utilize a watch that has both a digital stopwatch feature *and* a conventional face with hands, as shown on the right in figure 7–1. This will allow you to keep track of the immediate time of an exercise on the stopwatch, while looking at the big picture with the conventional face.

Fig. 7–1 The watch on the right is ideal for the assessment center as it has both the stopwatch and the conventional face.

Many candidates leave points on the floor by only utilizing a fraction of the time they are allowed. Many candidates utilize only three to five minutes when they are given 10 minutes to perform an oral presentation, for example. In this case, you should attempt to complete your exercise in around nine minutes. This will give you a minute of flexibility if necessary.

Follow the Directions

Another common pitfall for many candidates is that they do not follow the directions of the exercise. One candidate misread the directions of an oral presentation. He thought he had 10 minutes to prepare when he went into the room. He actually had 10 minutes to *give* the presentation. After sitting in front of the assessors for 10 minutes, he began speaking. They informed him that he was done and could leave the room. Ouch!

Do not let something so simple keep you from reaching your goal. Some exercises are complex, requiring the candidate to perform many tasks in a limited time. For example, an in-basket exercise may require you to give an oral presentation, submit written correspondence to the panel, state what you would ship up the chain of command, and to whom. If you forget about stating who you will contact in the chain of command above you, you will not receive credit and leave points on the floor.

One technique for following the directions well is to read them at least twice. The first time should be an overview. During the second reading, *highlight or underline the key points in the directions that require you to take action.* That way, you won't miss anything.

Another key point is to ask questions if you have any doubt. The proctors and sometimes the assessors can answer clarifying questions about an exercise. They will almost always ask if you have any questions. Even if they do not, ask questions if you have any doubt about the directions of an exercise. If they cannot tell you the answer to your question, they will let you know.

The emergency scene simulation shows the importance of this key point. If the exercise begins when you are dispatched, but you do not begin verbalizing your thoughts and actions until arrival (two minutes later), you may lose several minutes of critical time in which you could have scored points. Some candidates have remained silent during this time, leaving points on the floor. In contrast, if the exercise does not begin until you are "on scene" when the picture comes up, spend that time creating your tactical worksheet since verbalization is not necessary until later.

Attire, Appearance, and First Impression

Your attire is crucial as well. You must dress for success. A pressed suit with polished shoes and neat hair is the minimum to gain entry into the show. Like being on time, this is a must. A business suit with dark or neutral tones will express professionalism. Navy blue and dark gray are good choices. Men must wear a coat and tie. Ties should be darker in color, or red is okay as long as it's not neon. Women should wear a business suit, professional looking outfit, and /or a skirt of the appropriate length (just above the knee). Class A or B uniforms are common for promotion. Follow the directions given to you by your department.

In addition, lay all of your clothes out, try them on, and polish your shoes. Make sure your socks or hose and shoes match, and you have clean underwear. All of this may seem like overkill, but this effort will pay huge dividends come test day. Like a football team getting ready for the big game, the day before will be your pre-game mode. Ensuring all of these little things are in place will increase your confidence. How many other people have put this much thought into just the first impression? Not many.

Even a week before the assessment is a good time to try on clothes since you will then have time to get alterations made if needed. When you hang your clothes the day before, make sure the cat or dog don't use them as a sleeping bag that night. Carry a lint roller. The masking tape type works best.

Your hair should be cut above the collar, at least two days before the assessment so you have a chance to get used to it, and the tan lines can fill in. Female candidates should wear their hair up if it's long. Make sure you go to a trusted hair stylist or barber when you get your haircut. Two days before the assessment is not the time to try someone new. They may butcher your hair and your confidence will be shot. Go to your normal stylist and get your normal haircut. It's not the time to try some new look, unless of course your normal look is a mullet or hair past your shoulders (if you're a man). It's one less thing to worry about.

After you get your nice clean haircut, wash, fuel and detail your car the day before your assessment. There's something about getting a fresh haircut and driving your polished car. We tend to feel more confident, professional, and have a feeling of readiness to take on the world. If you're afraid your car will fall apart if you wash it, then borrow a friend's car. When you fill the tank with gas, detail it, shine the tires, and vacuum the inside.

Believe it or not, some panels look out the window to see the candidates arriving in the parking lot. If you step out of a clean car, on time, with confidence, they will already have a positive opinion of you before you even enter the room (see fig. 7–2). Contrast that with the rushed candidate, who screeched into the parking lot, jumped out of a dirty car, hair messy, throwing their coat on as they speed-walk into the building. Who do you want to be?

Fig. 7–2 This candidate looks sharp and is ready to go.

Panels don't necessarily care if are driving a Lexus, but what they will take note of is if you are poised when you arrive and if your car is clean. Remember, even if they don't see you, you will be more confident by virtue of the effort and professionalism you have displayed, which is the main purpose for doing this.

Another key ingredient to your confidence and first impression is to drive the route to your test the day before, at the same time of day as your scheduled assessment. You will then experience any traffic issues, road closures, or detours that may affect your travel time. You can then locate your parking spot and the building/floor to enter. All of this preparation will have you so confident that you will be like a caged tiger when your assessment comes. That energy will carry into the first impression.

A few other logistical issues must be handled the day before your test that will add to your preparation and confidence. Call a *trustworthy* friend and ask him/her to follow you to the assessment site. If your car breaks down on the way, simply pull your car off the road, lock it, and jump into your friend's car. Don't even think about the car. When you get the promotion, you can buy a new one. If the car doesn't break down, fine. Just have your friend drive right past the test location as you pull into the parking lot. Having this guardian angel will again add to your confidence, and remove Murphy's Law from your equation.

During my entry-level firefighter interview, years ago, a friend broke down just three miles from the interview location. He lost a fan belt. When he arrived at the interview 20 minutes late, the chief told him he could wash his hands and then leave. They wouldn't see him. Are you willing to add the worry of breaking down to your mind when you drive to an assessment? By removing this from your thoughts, you will be free to focus on your first impression and get your game face on.

Ask your friend to arrive from your point of departure 30 minutes before you want to leave. This will create a safety barrier if he/she is late. If they are 20 minutes late, they will actually be 10 minutes early.

Don't eat anything new, different, spicy, or weird-looking the night before the assessment. Also, don't eat or drink *anything* after you get dressed on assessment day. If you spill something on your clothes, your day will be shot.

Ask your trustworthy friend who's following you to the assessment to give you a wake-up call in the morning in case your alarm clock fails or the power goes out. Have that friend call you on your cell phone. If you have young children who will keep you up all night, then get childcare or check into a hotel. You *must* have a good night's sleep the night before your big day.

Developing Your Plan 

The goal of all of the sections of this book combined is that you will have an excellent understanding of what you will need to do to be an excellent officer *and* test like one in an assessment center. Then you will be able to formulate your own plan to accomplish these goals.

Hopefully by now you are starting to see the gap between where you are and where you want to be. The size of the gap does not matter. *Your effort and ability to make an effective plan, remain dedicated, and spend time are what matter.*

You have now been armed with the knowledge of what the KSAs are to be an excellent officer. It is up to you to develop them. You have also been educated on what an assessment center is and is not. You have been told what your mindset must be. Finally, you know what the types of exercises are on an assessment center, the key points for each, and how they evaluate the KSAs of a fire officer.

Your next task is to complete a self-assessment test to help you develop your plan. The test should help you determine your baseline: where you are. As you have been educated on the previously mentioned factors, you must now determine where you want to be. As you see where you are strong, plan to build upon those strengths. As you see where you are weak, plan to strengthen those weaknesses.

Some of the self-assessment questions do not have *right* answers. These particular questions are designed to promote thoughts and clarify your position on certain issues. Discuss your test with a mentor, your boss, or a respected officer. Ask them what *they* would do in each situation. Bounce the questions off several people to gain a consensus opinion and zero in on what *you* would do. Each of the self-assessment scenarios is from real-world experience. None of them are made up. After the questions, we've listed some thoughts for each question to help get you thinking. The thoughts are listed separately in order to let you come up with your own ideas prior to seeing what we did, sort of like the answers to a quiz. Once again, some of these don't have a *right* answer, but the Thoughts section will let you know what we did.

Combine the components of the barrier elimination, assessment center orientation, mentality, KSAs to be an excellent officer, exercises key points, and the following self-assessment test to create *your plan* (fig. 8–1).

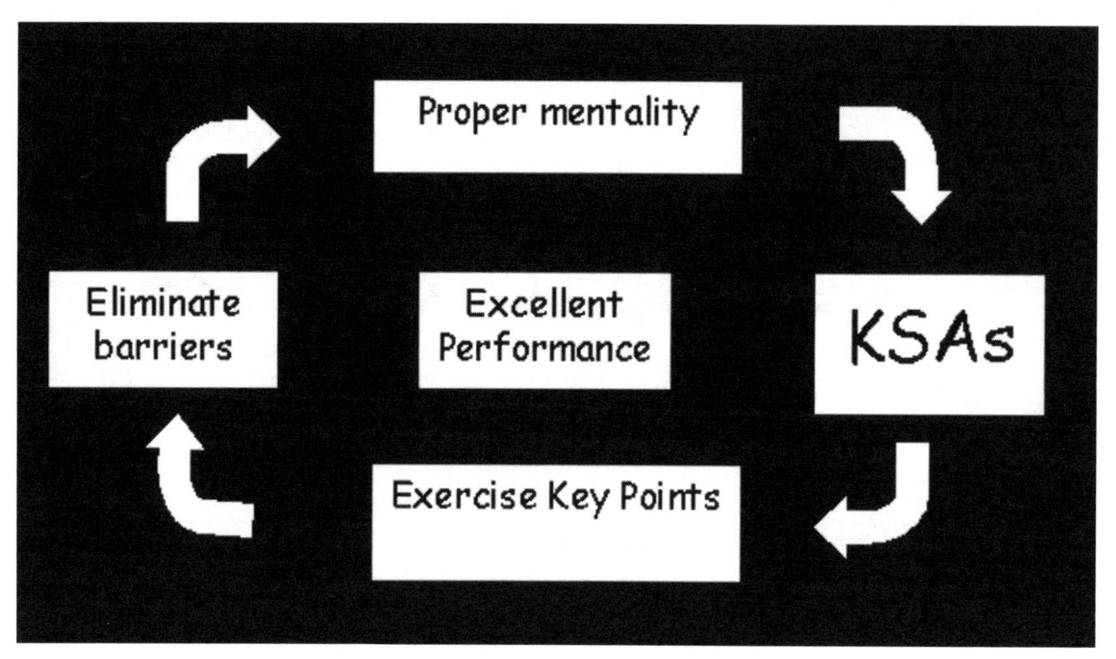

Fig. 8–1 Four components of a successful plan.

In conclusion, it is up to you to develop your behavior with the lessons you learn in this book. Remember, if you fail to plan, you plan to fail. No plan is worth the paper on which it's written *unless implemented. Through planning, hard work, and time, you will achieve your goals and dreams. There is simply no shortcut to success* (fig. 8–2).

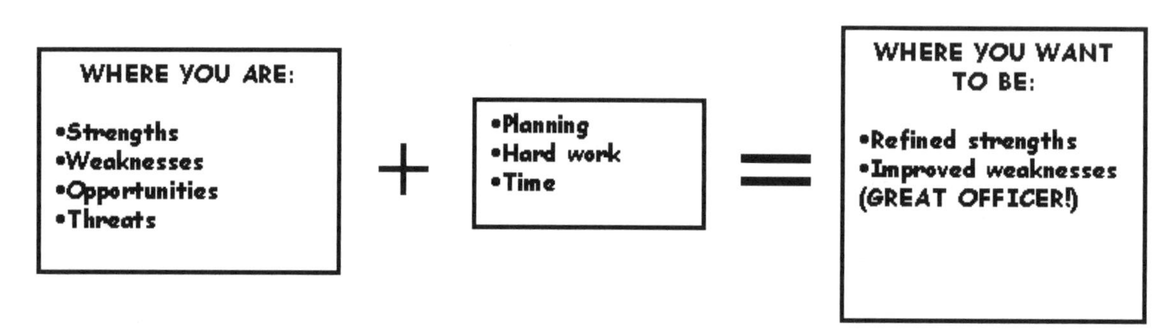

Fig. 8–2 The equation for success.

Transition from Doer to Leader

You may be aspiring to be a first-line supervisor (lieutenant, captain). This may be the first time you have been in a supervisory/leadership role. This is a big step for some of us. The transitional gap from doer to leader can vary from person to person. If you have never been in the leadership role, performing in an assessment center may be very uncomfortable.

The time to develop the KSAs is not a month before the test. You must make the transition from doer to leader as soon as possible. The longer amount of time utilized to develop your KSAs, the greater chance you have at being successful. Teaching classes, leading drills, performing public education presentations all take time. More importantly, becoming successful and comfortable with these leadership skills takes even more time. Do yourself a favor, start early!

Self-Assessment Test

1. When you are asked to give an oral presentation, how do you feel?
   a. Excited and look forward to the opportunity.
   b. Nervous and dread the idea.
   c. Neither excited nor nervous, but I will do my best.

2. When you are part of a group that has no leader, do you:
   a. Try to take the lead.
   b. Encourage someone else to take the lead.
   c. Wait to see what will happen.

3. How do you feel if you are put in charge of a project or activity?
   a. Happy that I was chosen.
   b. Unhappy that I was chosen.
   c. Neither happy nor unhappy, but I will do my best.

4. I want to be promoted primarily because:
   a. I want the raise.
   b. I want the power.
   c. I want the opportunity.

5. What is your first thought about State Fire Marshal officer classes?
   a. I like what I learn from the good ones.
   b. I need them to get a higher pay incentive.
   c. I need them to take promotional exams.

6. As an acting captain, you walk into a new firehouse and a seasoned firefighter (who's known for being difficult) is wearing Harley boots on duty. What would you do?
   a. Tell him to put on his work boots.
   b. Nothing, since I am an actor.
   c. Try to motivate him to take his boots off without being asked.

7. As a captain, you are teaching a probie how to start a circular saw. He is having a hard time, getting frustrated, and says, "I'll just get an axe!" What would you do?
   a. Start thinking about how do deal with his attitude.
   b. Keep working with him until he learns how to start it.
   c. Get a senior firefighter to teach him how to start it.

8. You are a BC. One of your engineers in your battalion who is approaching retirement has been losing his temper with customers lately and cussing at them. What would you do?
   a. Tell him to step into my office so we could discuss it.
   b. Tell his captain to talk to him about it, then see me.
   c. Nothing, he's done his time and is entitled to a little slack.

9. As a new captain, you notice that your 25-year engineer is not respecting you. What is a possible problem?
   a. He is an old salt and has a bad attitude.
   b. He tried promoting but was unsuccessful.
   c. He just needs to get to know you better.

10. How do you feel about morning meetings?
    a. They are a good way to get things organized in the morning.
    b. They are not necessary because everyone knows their job.
    c. They are almost impossible to hold due to calls.

11. What is the difference between leadership and management?
    a. You lead people but manage things.
    b. You lead those who want to be led and manage those who don't.
    c. Leadership and management are synonymous.

12. Excellent leaders are
    a. Born.
    b. Made.
    c. Both.

13. What is the best way to organize a written document?
    a. Intro, thesis, body, summary.
    b. Tell 'em what you're gonna tell 'em; tell 'em; then tell 'em what you told 'em.
    c. Write it in terms that the reader will understand.
    d. All of the above.

14. Motivation comes from where?
    a. The individual.
    b. The officer.
    c. The situation.

15. You are in charge of staging on a high-rise drill. Your first action would be to
    a. Assemble the companies by type.
    b. Separate the re-habs from fresh crews.
    c. Assign a scribe to assist you.

16. Who does the food unit leader report to (in ICS)?
    a. The logistics chief.
    b. The support branch.
    c. The service branch

17. You have an EMT ride-along. While on a call, the paramedic on the ambulance gets upset over having to transport a non-English-speaking patient. What would be your immediate reaction?
    a. Tell the paramedic to set a better example on scene.
    b. Separate the EMT student from the paramedic.
    c. Nothing, since the medic is just understandably tired.

18. What would be your sequence of operations upon arrival of a house fire?
    a. Give your arrival report, size-up, then direct your crew.
    b. Perform a size-up, arrival report, then direct your crew.
    c. Perform a size-up, direct your crew, then give a radio report.

19. Which of the following is a problem?
    a. A firefighter continuously late for work.
    b. An engineer who is rude to the new probie.
    c. A captain going through a divorce.

20. Which of the following is a symptom?
    a. A headache.
    b. A brain tumor.
    c. A hangover.

True or false?

21. I have spent more time thinking about the test than the position.
    a. True
    b. False

22. If possible, I would just take the job tomorrow and then figure out the rest as I go.
    a. True
    b. False

23. I have only taken classes that are required to take the test.
    a. True
    b. False

24. You can't study for an assessment center.
    a. True
    b. False

25. You can prepare for an assessment center.
    a. True
    b. False

Fill in the blanks.

26. I am great at (list job skills for positions I seek):
    a. ____________
    b. ____________
    c. ____________

27. I need some work in (list job skills for positions I seek):
    a. ____________
    b. ____________
    c. ____________

28. I am completely lost with (list job skills for positions I seek):
    a. ____________
    b. ____________
    c. ____________

29. Who are three of my mentors right now?
    a. ____________
    b. ____________
    c. ____________

30. The thing(s) I admire most about my favorite officer(s) is/are:
    a. ____________
    b. ____________
    c. ____________

31. The thing(s) I dislike most about my least favorite officer(s) is/are:
    a. ____________
    b. ____________
    c. ____________

32. I have taken ____ assessment center type tests.

33. I have had the most trouble with ________ type exercises.

34. I seem to excel in ________ type exercises.

35. Is there any correlation to questions 26–28 and 32–34?

36. I am involved in the following programs to develop my skills:
    a. ____________
    b. ____________
    c. ____________

37. I am taking these three classes to be a better officer:
    a. ____________
    b. ____________
    c. ____________

38. What can I do at work to become a better officer?

39. How long is my plan? What is the timeframe to accomplish my goals?

40. What are the strategic, tactical/objective, and task level goals for my plan?

Self-Assessment Test Thoughts

1. If you get nervous about oral presentations, eliminate the nerves by practicing as discussed in the oral presentation section.

2. Take the lead if you want to be a leader. Failure to take the lead is the biggest obstacle to effective leadership. By contrast, once you are an officer, look for opportunities to delegate and empower. Your reluctance to take the lead may be due to fear of failure or of others' opinions. Get over it. You will have to make unpopular decisions at some time in your career. Fear of failure will pass with time and with practice at just making good decisions.

3. Like taking the lead, take the opportunity to run projects to build your skills.

4. If you only want to promote for the money, here's a little secret—it's not worth it! Look for reasons like wanting to make positive change, wanting to take care of the troops, and other selfless decisions. Remember, it's not all about you.

5. If your State Fire Marshal classes leave a lot to be desired, look for other classes and opportunities like FDIC. Also, the instructor always makes the class. Ask around for the good ones. Taking classes *after* a test has been announced is usually too late. Take them way ahead of time to build a foundation.

6. The key here is that this dude is seasoned. He's an old salt. In addition, he's known for being difficult. He may just be baiting you, the *acting* captain, to start some conflict. Try to motivate him on his terms. In this case, I thanked him for being there and said, "Boy I'm glad you're here. This isn't my normal house and I'm just a guest. I'll be relying on your knowledge. How about if I cook dinner tonight?" His boots came off in about one minute *without* me asking.

7. If you get the senior firefighter involved, he is empowered and will take ownership (if he wants it) in the probie. Give the senior man a chance to show off his stuff to the new kid. You are then free to work on other projects.

8. Unfortunately, we have seasons in life where we lose our tolerance, patience, etc. The key here is that the captain should be the first point of contact to this engineer. You can gain insight from the captain, find out if anything is going on in the engineer's personal life, etc. Then coach and support the captain. If discipline is required, act on it, but not in a punitive way. Many of our salty old dogs feel forgotten, unappreciated, and retired before they actually retire. Help the captain help the engineer help himself.

9. Find out if he's tried to promote in the past. Perhaps you embody his frustrations. Sit down privately and get to know each other. While acknowledging his seniority and wisdom, offer to help him promote if that's what he wants. Use these traits to build upon and give him an opportunity to ride in your seat for training.

10. Do morning meetings whenever possible. They help you take your crew's vital signs in the morning and see how they are doing. You can get everyone on the same page. Some of the best meetings are over coffee at the tailboard.

11. We lead people and manage things. Things like hose, engines, trucks, and saws don't have emotions. People do. Leadership involves individual motivation. The key here is: don't treat people like stuff. They are more dynamic and more valuable. Engines don't get pissed off at you if you treat them poorly; people do.

12. You are made with certain strengths. Build upon those and work harder at the ones that need it. The key is, don't give up on the KSAs that don't come naturally.

13. All three are correct.

14. Motivation is internal, based on the individual. Everyone has his/her own set of motives. The key to the company officer is to know the motives (internal frequencies) of each member of your crew. Then you utilize that knowledge to inspire action.

15. I would get a scribe to help out. This gives you the help you will need while giving someone else an opportunity to learn.

16. Based on the ICS, the food unit leader reports directly to the service branch director, who then reports to the logistics section chief. The point of this question is that you should know *your* regional command system, whatever it is.

17. Separate the EMT student discretely. You must remove the student from the contamination of this paramedic. Once the paramedic notices that you did so, he will hopefully realize that something is wrong and start thinking about it. When he gets back to the station from the transport, ask him into your office and ask how the call went. Hopefully, he will cop to the attitude. If he does, ask him to apologize to the EMT and state that it was a bad example. He will take ownership and have an opportunity to fix the problem—which firefighters want—and that's what ended up happening here.

18. Size-up should come first to allow you to see what's going on. Then give your crew direction so they don't freelance since they are on scene and want to get to work. Finally, give your initial arrival report to the balance of the alarm.

19. A and B are symptoms, caused by an underlying problem you will need to investigate. The captain going through a divorce is a problem that can cause both impatience and being late.

20. The headache is a symptom that can be caused by either A or B.

21. Hopefully by now, you realize that you should focus more on position preparation than test obsession. While you need to pass the test to get the job, assessment centers cannot be faked. Get ready for the job and the test will take care of itself. That being said, overcome your obstacles and practice exercise key points so you do not become a great firefighter that "just doesn't test well."

22. While we will never know it all, be sure that you are diligent about learning. Remember, you will be responsible for your crew going home happy and safe at the end of the shift.

23. The minimum classes are never enough to be an excellent officer. If you want to get by with the least amount of effort, you will get people hurt.

24–25. Preparation for the assessment center involves position preparation. Part of that preparation is studying policy, procedures, tactics, and strategy, etc. Studying like you would study for a multiple-choice test is not enough. To approach the assessment center as if going in for an interview is also not enough. While you must research your job and department like you would to prepare for an interview, you cannot fake the skills that you must demonstrate in an assessment center. *Saying* you are good with people and *demonstrating* that are two very different things.

26–28. Ask around: Mentors, friends, co-workers, bosses, etc.

29. Who has taught you, taken time to nurture your skills, been patient when you blew it, etc.?

30. Look for the KSAs in your mentors.

31. Even if you do not like a certain officer, you may inherit some of his/her bad habits. Remember, most child abusers are formerly abused children. Do not repeat the cycle of dysfunction.

32. If you have failed them numerous times, have you focused too much on the test? Look at those obstacles and practice the KSAs. If you would make a great officer but can't test like it, then identify the barrier and attack it.

33–34. See if you can get your previous evaluation sheets or videos of your performance.

35. There probably is. There are no accidents. Be honest with yourself. Without significant changes, past behavior is the best indicator of future behavior.

36. Here's how to prepare for the position and shine in the assessment center: a) eliminate barriers, b) get involved to build KSAs over the long term, and c) practice the exercises.

37. Think of tactics and strategy classes, FDIC opportunities, leadership and management classes, college speech courses, etc.

38. The firehouse is full of opportunities with your crew—on calls, with projects, conducting drills, etc. Take advantage of on-the-job training opportunities.

39. The sooner you begin planning, the better. Do not wait for the test to be announced—it's too late by then. Plan two years out, at least. This doesn't mean you need to spend four hours a day for two years, but you should have a long-term plan to strengthen your skills. By now, you should know what you need to do. The further out you plan, the less daily impact to your schedule. You are spreading the load out over a longer term, thus reducing the daily impact to kids, family, work, etc.

40. Start broad in your goals, looking at the big picture. As you transition to the tactical/objective level, categorize your plan by barrier, weakness, strength, KSA, etc. Then, break each KSA down into specific tasks that you can perform to build the KSA.

Interviews 9

The following chapter will assist you with promotional interviews. Some assessment centers may have an interview component or exercise. This chapter will help you be more effective in any interview you might face.

The first section lists the 10 Interview Rules of Engagement. Then a short section on "making the connection" for promotional interviews is discussed. Finally, a sample interview score sheet and sample questions conclude the book.

Interview Rules of Engagement

We plan so much in life. We plan our meals, our chores, our retirement, and our finances (hopefully). We plan holiday meals, Super Bowl Sundays, and fishing trips. Many of us spend time thinking about where we want to live, who we want to marry, and what we want to do with the rest of our lives.

The funny thing is, when it comes to interviews, we don't plan much past "do the best I can and see what happens." Even though interviews are a crucial part of getting any job, whether flipping burgers or saving lives, few people plan the strategy of their interview ahead of time. No wonder that each day, thousands of people don't get the job they want. They simply didn't plan.

You are probably thinking, "How do I plan if I don't know what the questions are?" Well, the beautiful thing is that you don't have to know the questions ahead of time. Your plan will center on the 10 Interview Rules of Engagement. Later in this chapter, we have a short section on "making the connection" that will help you be even more effective during promotional interviews.

Since most firefighters love acronyms, we have one for the 10 Interview Rules of Engagement: BLACK SMOKE.

B	Blow them away with your first impression.
L	Look them in the eye.
A	Answer the questions directly.
C	Convey your strengths to the board.
K	Know yourself and your philosophies.
S	Speak from your heart.
M	Motivate them with your positive attitude.
O	Only be confident, not cocky.
K	Know the organization and job you want.
E	Easy does it, relax!

Think of the fire in your heart to promote. The fire is seen by the BLACK SMOKE in your interview.

Blow them away with your first impression

The old adage "you only get one chance to make a good first impression" is never more true and critical than in an interview. The panel will begin to score you before you even open your mouth.

During the first minute of your interview, your first impression will simply make or break your score. You absolutely must walk in smiling, happy to be there, with angels singing above your head, which is illuminated by rays of sunshine while a harp is heard in the distance. Figures 9–1 and 9–2 contrast the lousy first impression and the great first impression.

Your smile, body language, handshake, posture, enthusiasm, joy, energy, and respect are all sized up and determined by more than just what you say. The first impression will "set the table" for the interview. Do you want to use fine china and silver or plastic sporks and paper plates from your last visit to a fast food restaurant?

Although you have not answered a single question, the panel will begin to formulate their opinion (score) of you the moment you walk in the room. So how do you make a great first impression?

First, smile from ear to ear. The panel will see dozens or perhaps even hundreds of candidates. The panel will be tired, in a routine, and waiting to see someone that will light up the room. If you are early in the pack, you can set the bar high and they will remember you. You will be the standard from which all other candidates will aspire. If you are late in the pack, you can be that breath of fresh air that they are gasping for.

Fig. 9–1 A lousy first impression will be hard to recover from in the interview.

Fig. 9–2 A great first impression is critical to success.

Next, show your enthusiasm with your body language. Have a bounce in your step, excitement in your voice, and a firm handshake in your hand. Charisma comes not from what is said, but what is seen. It's that special glimmer in your eye that says "at last, I'm here and boy have I got something for you!"

Happiness is contagious. When you smile at someone, his/her natural human instinct is to smile back. By smiling at the panel, you show your true joy for having passed the written test and being one step closer to your dream. If the panel is happy because you are happy, you will successfully set the tone for the rest of your interview.

Be on time. In fact, be early. Most interview panels will not even see you if you are late. If they need to see 100 people, you just made their job easier when they eliminate you for your tardiness. Plan on being at the check-in location 10 minutes before your appointed time. This means that you must know the route, location, parking situation (including fees), and on what floor the interview is located. Having done so, these areas will be familiar to you on your interview day, which will help you relax and focus on your first impression.

The first impression is the *most* critical part of the interview. The panel will begin to score you in their minds and get a general impression of you from which you will be evaluated. You must blow them away. The first impression is much more than the first question. It's your smile, handshake, enthusiasm, body language, walk, and appearance. By acknowledging this, you can prepare from a week before up to the day before your interview. The right suit or Class A uniform, a clean haircut from a trusted barber, a detailed car, and a guardian angel will all add to your level of pride and confidence. That energy will transfer into the room with your first impression, and you will be starting off on the right foot.

Look them in the eye

Firefighting is based on trust. Trust is everything. Firefighters must trust each other with their lives, and the public has no choice but to trust us when we come into their homes and care for their loved ones. The interview panel will be looking for honesty and trust in you. You must be honest and sincere in your interview. The best way to show that what you are saying is true is to look the panel straight in the eye. The best answers in the world are absolutely worthless if you are perceived as insincere, lying, or not trustworthy. Like the first impression, trust is conveyed by actions, not words. The act of looking someone in the eyes while you speak says, "You can trust me because I am telling you the truth."

Even if you are being honest and just have the habit of not looking people in the eye, you will be perceived as being dishonest (see fig. 9–3). Have you ever talked to someone who has trouble looking you in the eye? What are they hiding? When kids lie, they look away. Another perception is that you are not confident if you can't look people in the eye (see fig. 9–4).

Nervousness can cause you to shy away from eye contact as well. But remember, shyness may be the reason, but the perception will be lack of honesty. If you overcome your difficulty with eye contact, you will score better on the interview. Just overcoming nervousness may help you with eye contact, and you will again be ahead of the pack.

First, let's talk about eye contact in the interview. The moment you walk in, as part of your first impression, look every single person in the eye and give him/her a strong handshake and a big smile. Do not ignore anyone. When you are asked a question, maintain eye contact with the person asking you the question. Begin your answer with the person who asked the question, then look at each of the other panel members in the eye as you answer. Conclude your answer with the same person who asked the question. Answer all the questions in the same manner—scanning, but maintaining good eye contact with each person as you go. They may not look back at you if they are writing notes. You may be looking at the tops of their heads. Just maintain eye contact with whomever looks back. The key is, do not ignore anyone.

Fig. 9–3 What message is this eye contact sending?

Fig. 9–4 This candidate seems much more confident and trustworthy.

Next, let's look at how to improve eye contact. Practice every day with everyone you meet. Also, practice with family and friends. Make a point to look them in the eye when you carry on a conversation. You may think you already do this, but chances are you don't do it as often as you think. Ask those closest to you and the people you speak to the most if you have good eye contact. If they can't remember, ask them to make a note of it for a few weeks so you can get honest feedback as you try to improve.

Another great habit to practice is to repeat and remember the names of the people you meet. How often do you meet someone and realize you forgot his/her name within five minutes? Pick a feature that will help you remember their name. Perhaps they look like someone you already know. Perhaps they look like a celebrity. Maybe you just need to repeat their name to yourself over and over again to make it stick. This will help your confidence. If you can remember the rank and/or last name of the panel members, that will make a great impression upon them. Coupled with eye contact, repeating a person's name conveys confidence, professionalism, sincerity, trust, and attention to detail; all qualities that officers must possess.

As you walk in and look the first panel member in the eye, your smile will light up the room. "Nice to meet you Chief Jones," will be your reply to the introduction. "I am honored to be here," will be your opening statement. Since you looked Chief Jones in the eye, she believes you. Since you remembered her name, she feels respected.

After the interview, you shake hands firmly again, looking her in the eye saying, "Thank you so much for your time Chief Jones, I hope to see you again." Chief Jones is thinking, *Wow, this is what I'm looking for: professional, respectful, confident, and I really feel like I could trust him.* If you can't remember Chief Jones' name at the end, don't freak out, and just refer to her as Chief. "Thank you so much for your time Chief, I hope to see you again, ma'am." She will feel just as favorable toward you.

If a panel member is not of rank, just use the term sir or ma'am if you can't remember their name. Never call them by the first name, even if you know them. If you can't remember their rank, look at the collar brass or badge to determine rank. Whatever you do, address them when you arrive and leave.

Looking the panel straight in the eye conveys trustworthiness, honesty, and confidence. Like the first impression, this act takes place before you answer any questions and will set the tone for a great interview. Coupled with repeating the person's name, looking panel members in the eye speaks volumes about you before you even open your mouth.

Answer the questions directly

"What color is the sky today?"

"It's beautiful today, but it looks like rain."

If that sounds weird to you, imagine what an interview panel would think. The question was not answered, which means the candidate does not listen very well. In the minds of the interview panel, this will translate into not following orders on the fireground, not listening during drills, and not hearing what is being asked. Since the answer does not have relevance to the question, the candidate will receive no points. In fact, the interview panel may be annoyed, but you made their job easier. You just won't get the job. Believe it or not, this happens all the time during interviews.

When you are asked a question—any question—you must answer it directly. You may then elaborate if needed. Let's look at an answer to the previous question that answers directly and then elaborates.

"What color is the sky today?"

"The sky is blue today, in fact it's very beautiful, but it looks like rain may come later." See the difference?

Let's look at another example.

"What are your short- and long-term goals?"

"I want to work for Dream Fire Department. Dream Fire Department has everything I have ever wanted. You have USAR, a HAZMAT team, and even a helicopter. What I like the most is that I will learn a lot about many different areas of the fire service."

What do you think? Good opening, but the answer tapered off pretty quickly. The candidate spoke more about what was so great about Dream Fire Department. He never said what his long-term goals were, and at best, just implied that the short-term goal was to work for Dream Fire Department. This happens all the time. Let's look at another answer to that same question.

"What are your short- and long-term goals?"

"My first short-term goal is to get promoted to captain by Dream Fire Department. Part of that goal is successfully passing my six-month probation and ensuring my crew goes home safe each shift. My next goal is to be an outstanding officer, learning all that I can, honing my style, and helping others reach their goals as well."

"My first long-term goal would be to join the USAR team. I feel that I can contribute my energy and enthusiasm to the team, which will help me to be an even better officer, an enduring goal of mine that will last my whole life. Finally, my personal long-term goal is to have children and raise them in a loving home."

This candidate nailed the answer! He answered it directly ("my first short-term goal is"), and then elaborated without forgetting the second part of the question ("my first long-term goal is"). Finally, the candidate elaborated by adding the personal aspect ("my personal long-term goal is"). This was a great, elaborate, direct answer to the question. Chances are they pinned a badge to his chest when he was done.

People fail to answer questions directly every day. It doesn't just happen in interviews. Have you ever asked someone, "How's it going?" Guaranteed, someone has answered, "Not much." You may even be confused as you read this. Here is another example. You ask a friend, "What's up?" He replies, "Pretty good."

See how easy it is? So what's the key to answering questions directly? First, listen to the question. Many candidates start to formulate an answer to the question before the panel is done asking it. The problem is that you will not hear the rest of the question because you are busy listening to your thoughts instead of the panel. Many questions come in two or more parts, which require you to listen to the whole thing before you can begin to answer. You *must* make an effort to remember the question as you answer. Questions are purposefully formatted this way to see if the candidate listens well.

Next, take a moment (about three seconds) to process the question and formulate your answer. Because of stress, many candidates will begin speaking the very instant that the panel gets done asking the question. If you have not taken a moment to process and formulate, you will begin speaking without really knowing what you want to say. You may even forget the question as you answer. Here's an example.

"Have you ever had to work in a stressful situation, and if so, what did you do?"

"Oh yes, I worked for a tough boss for three years who was an alcoholic. He lost his temper on a regular basis. He was often rude to the customers, which makes me glad to be a firefighter since we don't tolerate such behavior. In fact, the public has a right to proper customer service since they pay our salary."

The poor candidate totally forgot the second part of the question. He never stated what he did, which is really the most important part of the question. We have all had to work in stressful situations. The panel wants to know what you would do.

You do not have to know what the questions are ahead of time to be able to answer them directly in an interview. Simply listen to the question, take a moment to process and formulate an answer, and then directly answer it. Elaborate only after you have ensured that the question asked was definitely answered. A good rule of thumb is to try to answer all questions in the first sentence. Once that is done, you are free to elaborate as needed to give depth to your answer. Practice with friends and relatives and ask them how you are doing. Are you answering what was asked?

Taking the time to listen, process, and formulate will allow you to answer *any* question that will be asked. Simply slowing down will give you the chance to relax, think, and organize your thoughts. Most people are nervous because they feel like they have a gun to their head when they answer questions in an interview. The fact is that you have plenty of time to organize your thoughts, which will relieve some nervousness as well.

Think of it this way. What if you had half an hour to answer one question? You would think, "Wow, plenty of time, no stress." Well, three to five seconds feels like half an hour in an interview. We hate silence in an interview. In fact, we hate silence in ordinary everyday conversation. That is why people say some pretty stupid things. Everyone hates the emptiness of silence and makes conversation to fill it. The same holds true in an interview. Since we hate silence, our nerves cause us to rush to answer a question without really knowing what we want to say. The result is that we spend about three or four sentences blabbing until we find an answer that we want. By that time, we're lucky if we remember the question in the first place, and we fail to answer the question directly. It happens every day.

Answering questions directly is crucial to successful interviewing. If you fail to answer the question asked, the panel will score you poorly. You will convey a sense of poor listening skills and inability to follow directions. You must be skilled at both of these traits, which are vital to good officers.

Lack of answering questions directly happens everyday in casual conversation. Practice listening to questions with friends and family, taking the time to process and formulate an answer, and then answer directly. Elaborate only after you have answered the question asked. Since people generally hate silence during a conversation, the nervousness in an interview creates a need to instantly answer the question without taking the time to process and formulate. Simply taking three to five seconds to slow down and organize your thoughts will pay huge dividends, and allow you to answer questions directly.

Convey your strengths to the board

If you asked a car salesman to give you the best features of a car and he stood there with a silly look on his face, unable to answer, would you buy the car? Probably not.

Very few people have a good sense of their strengths. When practicing interviews, 90% of candidates asked about their strengths have to stop and think about it (see fig. 9–5). This shows a complete lack of preparation, confidence, motivation, desire, and goals. Your strengths must be as familiar to you as your name, address, phone number, and favorite ice cream.

A good car salesman knows all of the options on the car he is selling. He is ready to answer any question about the gas mileage, horsepower, warranty, and extras that the car has. You must do the same with your strengths.

Fig. 9–5 You should not have to work this hard at thinking of your strengths.

You must also have reasons for and examples of your strengths. Just knowing your strengths is not enough. You must have concrete examples for each trait. For example, if you say that you are a great team player, elaborate by discussing your experience coaching a football team or being in the Kiwanis. Give specific examples within one of those. If you state that you are a great team player by being a member of the Kiwanis, give examples of committees, projects, or meetings in which you displayed team work. Only tell them the truth. Remember, you will be looking them in the eye.

As you paint a picture of each of your strengths, the panel will be superimposing you into the role of officer. They will see your strengths as an asset to the department. We will discuss how to make that connection in a later section of this chapter. For now, think of five strengths and have at least one example per strength. Even if the board doesn't ask you to specifically describe your strengths, having them in mind when you arrive will pay huge dividends during the interview.

For example, the panel asks, "How have you prepared for this job?" If you know your strengths you could answer, "I have prepared for the position of captain with Dream Fire Department by honing my strengths of teamwork, dependability, and integrity. First, since firefighting is a team sport, I have strived to participate in team activities at every opportunity. A good example is my work as a member of our apparatus committee. We routinely work on apparatus specifications that meet the goals of our department. To meet those goals, I have communicated with our entire membership across ranks, divisions, and shifts so that everyone has an opportunity to have input. Second, my integrity and dependability has been seen in my involvement in the chaplaincy program. I maintain professionalism and remain vigilant to respond to the needs of our team members and the public. I must remain sensitive to and empathetic to make a positive difference. These skills that I have honed will transfer to my crew as a captain. I will be a consistent team builder and will remain dependable with the utmost integrity."

By answering in this fashion, you have first answered the question directly, and then you weaved your strengths into the answer when you elaborated for depth. Remember, the panel does not have to ask, "What are your strengths?" for you to convey them during an interview.

Conversely, do not force your strengths into questions that are not appropriate. For example, the panel asks, "What would you do if a member of your crew arrived to work with the smell of alcohol on his breath?" Do not say, "Since one of my strengths is honesty, I would immediately tell the battalion chief that Joe is drunk. It's better to be safe, than sorry. Safety is another one of my strengths. I am always safe . . . safe, safe, safe . . . that's me."

See the difference? You sold out Joe in order to toot your own horn. Turns out that Joe is diabetic. He just had a bit of trouble with his blood sugar level. You demonstrated your *lack* of teamwork by assuming the worst in Joe, and your poor problem-solving skills by failing to realize that the smell was acetone.

Once again, only convey your strengths when it is appropriate. Many other questions are designed to assess your problem-solving skills, communication skills, thought process, etc. You can show your strengths by answering questions effectively, even though you do not specifically cite your strengths.

Let's look at that question again. "What would you do if a member of your crew arrived to work with the smell of alcohol on his breath?" A much more effective answer would be, "First, I would talk to the firefighter. I would tell him that I smell alcohol on him and that I want to make sure he's okay. That way I don't assume the worst, but I am checking out the situation because our safety may be compromised. Diabetes can be mistaken for alcohol sometimes, which could be the case here. In fact, low blood sugar can affect job performance. If I believed that it was alcohol, then he is obviously not fit for duty. Either way, the battalion chief must be notified since he will need to make a staffing adjustment. If this is indeed alcohol, then he has violated our policy and I must document the issue to ensure proper disciplinary action is taken. I would also ensure he had a safe way home from work."

In this answer, you showed your teamwork ("I don't assume the worst"), you showed your problem-solving skills ("I am checking out the situation"), you showed your attention to safety ("our safety may be compromised"), and you showed your respect for the chain of command ("the battalion chief must be notified").

In addition to your strengths, think of one or two weaknesses. We all have them. To say that you have no weaknesses is arrogant and ignorant. First, be honest in your answer and then answer directly.

You have two options when asked about a weakness. First, you can discuss how you are attempting to improve a trait: "Public speaking is a weakness of mine, but I am attending a speech class to help me improve." A second option is to discuss how your weakness is often an asset: "I am pretty impatient. When I want to accomplish a goal, I work very hard to get it done. I never forsake quality, I just work that much harder to get it done." Always end on a positive note.

Do not say, "I am a perfectionist." This is a canned answer, and the panel will most likely roll their eyes at you. Be honest. If you state that you are a perfectionist, you are again stating that you have no weakness or that you are trying to be perfect.

Remember to *convey* your strengths to the board. The rule does not say, "State your strengths." Convey simply means that you send the message about your strengths, either directly or indirectly, through the questions that are asked. As you have seen by the previous examples, you can convey your strengths without specifically stating them. The panel does not have to ask, "What are your strengths?" to give you the opportunity to convey them. When you do convey them, be careful to do so at the appropriate time and in the appropriate way. Do not force it. The previous examples should be a good reminder of the difference.

Conveying your strengths to the panel in an interview is as important as knowing the product you want to sell to a customer. You are there to sell yourself. You cannot possibly do that without knowing your strengths, conveying them, and giving specific examples. Giving specific examples lends credibility to your statements and helps the panel understand how you will be an asset to the fire department. Since you do not know what the panel will ask, you must attempt to weave your strengths into your answers. Be careful not to force it. You may not get a black-and-white question about your strengths. Questions are often designed to look for specific traits. Answer the question directly. You may have to convey your strengths through your answer without specifically stating them.

Know yourself and your philosophies

"Know your self and your philosophies" probably sounds a lot like "convey your strengths." There is a difference. Unlike your strengths, you have definitions, opinions, thoughts, experiences, and philosophies that have molded your character. Your definition of customer service, sexual harassment, or the chain of command has little to do with your strengths. Why you want to be a lieutenant, get a promotion, what you have done to prepare for the job, and your goals are also different from your strengths.

What *does* make you want to be a lieutenant, captain, battalion chief, or higher? Perhaps family tradition, a unique experience as a child, or a friend has influenced you. When have you overcome adversity? The list goes on. Although you may want to tie these into your strengths, knowing yourself and your philosophies will require you to do some homework.

Not knowing yourself is like using a pencil to paint a masterpiece. You must use all the colors available to paint the most beautiful and accurate picture possible. The more you know about yourself, the more colors you have to paint with.

As you will see later, you must do homework in three subjects: yourself, the fire department, and the job you seek. Whether your goal is lieutenant, captain, or battalion chief, you must have done your homework in these three areas. For now, we will focus on you.

The panel will ask two types of questions: open-ended and closed-ended. Open-ended questions let you take the question where you want to go. These are often indirectly answered. For example, the panel asks, "tell us about yourself." You could go in a number of different directions. You could speak of your strengths; tell why you want the job, state why you are proud of Dream Fire Department, etc. You have much more latitude in open-ended questions.

Closed-ended questions are a bit more difficult to answer. They require very direct answers and less latitude. For example, "Define sexual harassment." There is not a lot of room for interpretation. You either know it or you don't. Another example would be, "What is the chain of command?"

The key is to find your definition and philosophy in areas affecting firefighters and fire departments. The panel does not necessarily want a textbook definition to these questions, they want *your* definition. Once again, they are trying to get to know you, how you think, and who you are.

So how do you do your homework on you? Well, simply think about the following topics. While reviewing the checklist that follows, ask yourself three things:

1. What is my definition of this?
2. What is my experience (good and bad) with this?
3. What is my philosophy or opinion of this?

If you have no idea, *now* is the time to ask others—before your interview. Firefighters, friends, teachers, parents, and books all have the information. You are the only one who can decide for yourself after you gather the knowledge. Some items do have only one definition, but others are open to interpretation. All will require you to think. Take notes and mark the items off as you go.

Integrity	Character	Motivation	Ethics
Trust	Honesty	Allegiance	Culture
Relationship	Respect	Empathy	Leadership
Communication	Forward thinking	Customer service	Retirement
Internal customer service	Public opinion	Brotherhood	Sisterhood
Fairness	Equality	Equity	Micromanagement
EMS	Family	Free time	4-days
48/96	Shift work	40-hour week	Staff
Line	Administration	ICS	Tradition
Progress	Structure	Semi-militaristic	Chain of command
Following orders	Sexual harassment	Hostile work environment	Discrimination
Fairness	Disparate treatment	Policy	Procedure
SOP	Transporting patients	Weapons of Mass Destruction (WMD)	HAZMAT
USAR	Wild land	Airport Rescue & Firefighting	Aviation
Special ops	Downsizing	Adversity	Failure
Judgment	Ridicule	Mistakes	Bad experience
Bad boss	Strength	Weakness	Success
Credit	Experience	Education	Unions
Budgeting	Seniority	Pay incentives	Contracts
Binding arbitration	Neutral	Safety	Retirement
Goals	Fire engine	Fire truck	Company
Chief	Battalion Chief	Captain	Engineer
Firefighter	Volunteer	Paid	Professional
Performance	Job	Lifestyle	Heart
Head	Flexible	Difficult	Stern
Fun	Work	Play	Joy
Happiness	Sorrow	Death	Pain

Anger	Denial	Love	Hate
Fire	Loss	Pleasure	Probation
Fire academy	Teamwork	Togetherness	Training
Reading	Studying	Learning	Sleeplessness
Stress	New York	9/11	Pearl Harbor
Terrorism	War	Peace	Frustration
Veterans	Medicare	Welfare	Society
Politics	Ice cream		

Obviously, some heavy items were on the list. That's why we ended with ice cream—to sweeten the list. Chances are there were several items that you never associated with firefighting. The fact is firefighting has become a complex profession requiring depth from the people who seek to become officers. You *will* be asked questions about these topics. You must think of these issues and others so you have a better understanding of where you are and where you want to be. The effort made on self-study will show in an interview. Using our car analogy again, how do you come from the factory? What makes you go? What are your definitions, experiences, and philosophies in life?

Take the time to make notes and outline your package. Also, ask yourself questions as you think of the items on the list. Here's a list of examples:

- Why do I want to be an officer?
- Who has influenced me?
- What was a tough, stressful, or painful experience I have had?
- How did I overcome it?
- What have I done to prepare for the job of lieutenant, captain, or chief officer?
- What are my short- and long-term goals?
- What does integrity mean to me?
- When have I given great customer service?
- What is my leadership style?

As you can see, it's very easy to get quite a list going just from the items in this section. Very few people will go to this effort to prepare. The gap between you and your goal is narrowing, and you are still ahead of the pack.

In addition to your strengths, you must know yourself and your philosophies. The interview panel wants to get to know you in a very short time. Going into an interview to sell yourself without knowing who you are is like trying to paint a Picasso with a pencil. The more you understand who you are, the better prepared you will be to paint a beautiful picture with the most colors possible.

Firefighting today requires much of those who choose the profession. Taking the time to review these topics will give you a chance to do your homework on you. No one else can do this for you. The depth of knowledge you have regarding your definitions, experiences, and philosophies in life will arm you with a vast array of colors from which to paint your portrait. Go paint a masterpiece!

Speak from your heart

Most people worry about what they think the panel wants to hear. Let's clear it up once and for all: *speak from your heart.*

The panel wants to hear the truth, which is why we devoted a whole section on looking them in the eye. Looking them in the eye does not amount to a hill of beans if you are not honest. You must speak straight from the heart—being completely honest—throughout the entire interview. Remember, the panel wants to get to know you, the *real* you. They do not want some puppet that has been coached to give canned answers in hopes of tricking the panel into liking you. They will see right through it.

Another reason to speak from your heart is that it's easy and calms your nerves. Just say what you feel. Be yourself. Do not try to talk in some way that is not normal for you. Most candidates get very nervous because they are looking for the "right answer" to the questions. This creates unnecessary anxiety. They want to know what the panel is looking for and what they want to hear. Well, unless you have ESP, that's impossible. Once again, they want to see and hear the real you. If they feel they are being fed a line, they will score you low. Trust, honesty, and integrity are the foundation of any good candidate.

Now, if you peed your pants on the way to the interview, you can hold that one to yourself. However, if they ask you why you want to be a captain, tell them the truth. There's a difference between, "I like helping people," and, "Being a captain for Dream Fire Department would be a dream come true for three reasons: First, I want to be in a position where I can lead our firefighters. I have the ability to motivate, inspire, and help people reach their goals and get the best out of themselves. I love doing that. Second, I want to be the one responsible for getting our troops through tough situations. I am good at remaining calm when things are chaotic. I like building a team that can endure long shifts and tough calls together. Finally, ever since I was little, I dreamed of being a firefighter. I don't know how it started, but it hasn't ever gone away. I want it more every day. I eat, sleep, and dream of the fire service. To be a captain in our department will truly be a dream come true." See the difference? *That's* speaking from the heart.

Speaking from your heart is where you will get the passion to be motivated and energized. You want the panel to see your passion and desire. Your heart is where your passion lies. Let honest desire for the promotion come through.

Speak from your heart. Tell the truth. Like looking them in the eye, the panel wants to see that you are being honest. They want to see and hear the real you so they can make an accurate assessment of your ability to do the job. Speaking from your heart is impossible to fake. When you take the time to listen to the question, process and formulate your thoughts, and look them in the eye, speaking from your heart will be easy, fun, and give them what they want—the *real* you.

Motivate them with your positive attitude

If honesty and trustworthiness are the top qualities of a good officer, then a positive attitude is a close second. Attitude is simply how we respond to what happens to us in life. Attitude is the only thing we have complete control over in life. We cannot control the weather, other people, or even our health. Many joggers have been hit by cars or dropped dead due to a heart attack. How much less control do we have as firefighters? You can expect loss, pain, sadness, and sleepless nights. When days like these come along, we all want to be around someone with a healthy, positive attitude (see fig. 9–6).

Fig. 9–6 The highly motivated candidate.

The most respected officers are not the ones with the highest rank, most training, or biggest biceps. The best officers are the ones that always have a smile, are willing to help, have a nice thing to say, and a great attitude to share. They cook for the crew. They make everyone laugh. They help out everywhere and put others first. Remember the saying, "No one cares how much you know, until they know how much you care."

Happiness is contagious. When we are around people with high energy who are happy and motivated, it rubs off on us. As we mentioned earlier, when you smile at someone, his/her instinct is to smile back at you. Happiness and enthusiasm work the same way. When you walk into the room, you want it to light up with your pure joy, enthusiasm, and happiness to be there.

We all want to work with and be around people who bring us up, improve our attitude, and make us happy. Firefighting is no exception. In fact, a great attitude is even more critical to firefighters since we live together, and our job is incredibly demanding. Firefighters are like a family. We share some of the best and worst times in life. We rejoice together when we save a life and comfort each other when we lose a life. We share time off together with our families, and our kids grow up together.

The best way to show your positive attitude is just that—*show* it. Remember your first impression? That is the time to walk in and share your energy. Your smile must be sincere and light up the room. Shake hands firmly and thank the panel for the mere privilege to interview for your dream job.

When answering questions, weave your positive attitude into what you say, and *how* you say it. Smile as often as possible. Show an almost child-like motivation and desire to get the job. If you have prepared by knowing yourself, your philosophies, and your strengths, then you will have a blast.

Let's look at an example. "Why do feel you are the most qualified for this job?"

"I'm so glad you asked! First, I am the most motivated person you will see today. I believe that motivation and a positive attitude are crucial to the rank of captain. My passion for the fire service comes from a long family tradition. Firefighting is in my blood, which means that I will work as hard as anyone to make Dream Fire Department the best it can be. I want Dream Fire Department to be better for me being an officer in it, and I think that my love for the job will make that happen. I will always set a positive example. Second, I love teamwork and am good at building teams. I've been a football coach, a member of the apparatus committee, and music team member at church. In each of these teams, I put others first, helped out whenever possible, and only measured my success by my team's success."

This person is pretty fired up. In my entry-level interview, I looked the panel right in the eye, with a big smile and said, "I am the most motivated person you will see today. I am the one you are looking for. I eat, sleep, and breathe the fire service." Some years later, I saw my score sheet from that interview. One of the panel members wrote, "The most motivated person I have ever seen."

Although you want to maintain a professional and composed manner, you also want to jump out of your skin with positive energy. Have fun in the interview. Speak from your heart and go get your promotion.

Your energy, enthusiasm, motivation, and positive attitude must shine through in your actions and your answers during an interview. The panel will immediately be inspired and motivated by your childlike love of the job and desire for the promotion. They will remember what it was like when they were in your shoes. By starting with the first impression and weaving your positive attitude into your answers, the panel will see that you *are* the one they are looking for.

Only be confident, not cocky

Firefighting is a team sport; more like football than golf. None of us makes it alone in this business. We are all products of the hard work, love, and support of our family, friends, and fellow team members. On the job, teamwork is the bedrock of all that we do. A good officer will not take all the credit for anything good that happens. Success in the fire service is based on teamwork. Thinking you know it all, did it all, been there, and done that all by yourself is cocky. The fire service has no room for cockiness.

Figure 9–7 shows the contrast between the confident and cocky candidate. These are both captains aspiring to promote to BC. The BC candidate on the left is confident and motivated. The BC candidate on the right has a "been there done that" attitude with his body language. Who would you want to promote?

We have spoken a lot about being confident so far. Unfortunately, confidence can easily be perceived as cockiness. Cockiness is not appropriate in an interview. Cockiness turns people off. The interview panel is no exception. There is a fine line between the two. Cockiness says, "I am perfect, and you would be wrong to pass me by." Confidence says, "If you give me a chance, I will prove I am the right person for the job." Do you see the difference?

The key to avoiding cockiness is to remain humble and respectful of the panel and the job. You can convey your strengths, exude confidence, and be highly motivated while maintaining a respectful and humble attitude.

Think of this equation: *Motivation + respect + humility = confidence.*

Fig. 9–7 The cocky candidate on the right is easy to spot.

Let's look at some sentences that contrast between confident and cocky. In each of these pairs of examples, the first sentence is confident, and the second sentence is cocky:

"I am the most motivated person you will see today."

"I am better than anyone."

"If given the chance, I will make you proud you promoted me."

"If you don't hire me, you will make a big mistake."

"I learn from my mistakes."

"I don't make mistakes that often."

"Teamwork is one of my strengths."

"I am usually the best person on the teams I belong to."

"If you could talk to the people I have worked with, you would see that I am hard working and dependable."

"Everyone I know would tell you to promote me."

"I would love to be a battalion chief for Dream Fire Department to learn and contribute."

"I should be a battalion chief for Dream Fire Department to show you what I know."

Notice the differences in the sentences. The main difference is respect and humility—respect for the position you seek and the interview panel, and humility in showing that you do not know it all. They are both crucial. You must strike a balance between humility, confidence, motivation, and respect. Remember, no one makes it alone or knows it all in life, much less in the fire service.

During an interview, you must be confident without crossing the line into cockiness. The key ingredients to preventing cockiness are respect and humility. Being a fire officer requires teamwork and selflessness. Teamwork in the fire service is based on putting others first and accomplishing a common goal. None of us makes it alone. Remember that, and you will have a healthy sense of confidence without being cocky.

Know the organization and job you want

Remember the earlier section on knowing yourself and your philosophies? We mentioned that in addition to you, homework would need to be done in two additional areas: the organization and the job.

You cannot possibly articulate your desire to promote in an organization you do not know or have a job that you do not understand. You must do tremendous research in both of these areas. Just as knowing your strengths and your philosophies are keys to selling yourself in an interview, knowing the fire department and job will enable you to paint a more accurate picture of your *fit* into the position you desire. You again will be able to paint with many colors as you discuss your particular fit to a fire department's leadership culture, training philosophy, or value system.

In like fashion, you can show your understanding of the job you seek, whether lieutenant, captain, battalion chief, etc. By doing your homework, you will arm yourself with knowledge of the responsibilities and KSAs to be excellent at the job you want.

Create a formula for combining your knowledge of yourself, the organization, and the job that will allow you to provide the interview panel an excellent sampling of your ability to do the job you desire. For now, we will focus on the homework portion of the organization and the job. As in the earlier section on you, we have some topics that you should know about regarding the organization and the job. Many are the same; however, we will look at them from a question standpoint. If you do not know the answer, find it. Talk to someone who has the job you want, attempt to do a ride-along with a friend, etc. Ask yourself each of the following questions:

The organization:

- Why do I want to work for Dream Fire Department in an officer capacity?
- What is the population of the jurisdiction?
- How many stations do we have?
- How many engines, trucks, battalions, and ambulances do we have?
- How many people are in the organization?
- What is the organizational structure?
- What are the divisions (operations, prevention, EMS, training, etc.)?
- What is the chain of command?
- What is the culture?
- What is the annual budget?
- How is the budget funded, organized, spent, allocated?
- Are we aggressive about training?

- Do we have volunteers?
- What are the current hot topics?
- What are the current goals?
- How can I help the department meet those goals?
- What is the current labor/management relationship?
- Are we growing?
- Do we have transporting ambulances/paramedics?
- What are our special operations (USAR, HAZMAT, aviation, etc.)?
- Do I have an excellent grasp of our SOPs?
- Do I know the labor contract well?
- Can I see myself retire from Dream Fire Department?
- Do I want to spend 25–30 years at Dream Fire Department?
- What are the current challenges facing this department, and how can I help the department meet them in my new job?

Believe it or not, most candidates would not be able to answer most of these questions, even for their own departments. They just know that they want a job or promotion. We live in a very transitional society in which promises are long and commitments are short. Fire departments invest a lot of time and money into new recruits and officers. They want to know that you plan to be around for a while since you have *specific* reasons that you want to work for *their* organization.

Now, let's look at some questions about the job. Once again, you may have to hunt around for the answers, but the time will be well worth it. If you are seeking promotion, spend a lot of time with people who are great at the job you want. Learn from them and emulate them. Experience is the best teacher. Gain from their wisdom and knowledge. We should all have a mentor. Pick people who are respected by the folks with whom they work.

The job:

- Why do I want to be a lieutenant, captain, battalion chief, etc?
- What specifically about the job appeals to me?
- Why would I be good at it?
- Who do I know that is good at this job?
- Who has been a mentor to me?
- When did I first know about this job and realize I wanted it?
- What does it take to be great at this?
- What KSAs do I need to possess?
- What is different between what I need to know to get the job and then be great at the job in the years to come?
- What are the responsibilities?
- What would be my short-, medium-, and long-term goals when I get this job?
- Who would I report to?
- Would I be part of a team or alone most of the time?

- What are the tools used to do this job?
- What is the training needed to do this job well?
- How much training is involved after I get the job?
- Where would I go or with whom would I talk to find more information or get more training?
- How would this job affect my family and time?
- How would this job affect me emotionally, mentally, and physically?
- What is the same as what I do now?
- What is different than what I do now?
- How can I build upon what I have to be great at this job?

As you can see, the list of questions to consider about the organization and the job is quite extensive. Take the time to make notes. Do not simply gloss over these questions. You must take the time to explore these areas. You cannot impress a panel without this critical information.

Like the homework required on you, homework on the organization and job is critical. Most candidates fail to do the work up-front, and it shows in an interview. As you gain knowledge and perspective, the work you have done to organize your thoughts will show an interview panel that you are painting with all of the colors of the rainbow.

Easy does it, relax!

We have finally made it to the last of the BLACK SMOKE Interview Rules of Engagement: Easy does it, relax. As you have seen, none of the rules require you to know the questions ahead of time to excel in an interview. All of the rules apply to any interview for any job in the fire service. Our final rule is no exception, but it is perhaps the toughest to do.

If you are relaxed in an interview, you will be able to perform and take advantage of all of the rules you have learned. You will have a great level of awareness and utilize *all* the Rules of Engagement. You will make an awesome first impression, remembering to look the panel in the eye. Your composure will show your confidence, professionalism, and maturity. Your relaxed state of mind will help you listen to and understand the questions, answering them directly. You will easily weave your strengths and philosophy into the answers. As you speak from your heart, you will have fun, showing your enthusiasm and motivation. You will have a wonderful interview.

Most people believe that they have no control over their nerves in an interview. Public speaking is the number one fear of Americans. Some would actually prefer death to making a speech or conducting an interview. The fact is you have more control than you realize. Refer back to chapter 1 for information about removing your barriers.

While we all get a bit nervous when public speaking, three key principles will help you control your nerves.

The first key to controlling the level of nerves is *preparation*. The more you are prepared in any given situation, the less nervous you will be. How many times have you either been given a pop quiz or taken a test that you were not prepared for? If you had not studied, chances are you were a basket case. I still have nightmares from college. The nightmare is always the same. I am a nervous wreck because I forgot to study for a test or missed class when the information was given.

Conversely, when we are prepared, confident, and eager to show our stuff, we are much less nervous. The whole point of the 10 BLACK SMOKE Interview Rules of Engagement is preparation. Live by these rules and you will be prepared. Be prepared and you will be eager and relaxed. That is what you want for your interview. Be eager to show your stuff.

The second key to controlling your nerves is to realize that *you don't have what you already ain't got.* A wise old battalion chief used that saying to tell people not to sweat interviews. His point was that you have nothing to lose, because you do not have the job in the first place. The panel is not interviewing you to *keep* your job. You do not have it yet, so you cannot lose it. This sobering reminder actually takes the pressure off you. The panel is not going to take your first-born child or cut off your pinkie toe if you do poorly in the interview. They are not going to hold a gun to your head as they listen for the "right" answer.

Remember, they just want to get to know you, the *real* you. The right answer is the one from your heart.

Many candidates do not get their "dream" job, only to get a better one the next time. I once coached a firefighter candidate who was so excited because he aced his interview and only had to pass a psychological exam to get his "dream" job. No problem, right? After all, how do you study for a psychological exam?

Well, he failed the psych exam. He was understandably distraught. He lost his dream job and felt that the failure of a psychological test would haunt him at every subsequent fire department for which he tested. Two weeks later, he was hired by his hometown department, which is four times the size and pays almost twice the salary!

The third key to being relaxed is to remember that *the panel is on your side.* The interview panel wants you to succeed. They want to see you at your best and understand that interviewing is not easy. They are eager to see highly qualified candidates who will improve their fire department. In addition, they have all been in your shoes before. Everyone who serves on an interview panel has been interviewed, probably more than once. They are not the enemy, or big and bad scary people. They are truly pulling for you. That being said, you may see a panel member who does not seem as nice or friendly. Do not worry. Sometimes certain panel members just seem less friendly than others. Chances are that panel member will score you higher when you reflect the 10 Rules of Engagement.

All of the previous BLACK SMOKE Rules of Engagement have required preparation, thought, and planning on your part. By remaining relaxed, you will have the presence of mind to fulfill all 10 of the rules. Your interviews will subsequently improve dramatically. Excess nerves will cause you to forget the Rules of Engagement. As you control your nerves through preparation, remembering that you have nothing to lose, and that the panel is on your side, you will be relaxed and fulfill your plan and more importantly—fulfill your dream!

Making the Connection

Making the connection is the key to promotional interviews. You must make the connection between yourself, the job/position, and the fire department. Show the interview panel the conduit between these three components. By answering the questions about yourself, the job, and the fire department as outlined in the BLACK SMOKE section, you will be armed with the knowledge to show the panel your stuff.

Making the connection is the art of elaborating during an interview. It gives depth to your answers instead of reciting your resume. By keeping this as a priority, you can have a much more effective interview. Let's look at some examples.

The panel asks, "What are your strengths?" A simple answer would be, "I am good with people and have a lot of common sense."

An answer that makes the connection would say, "First, I am excellent with people. As you know, a fire captain must have the ability to motivate, inspire, and lead. I have done all of these in my role as a manager with our USAR team. That experience has given me ample opportunity to motivate my team, get results, and keep them safe in the process. Second, I have common sense. Common sense is required of Dream Fire Department officers. We must be able to see the big picture while our firefighters, paramedics, HAZMAT technicians and other specialists are doing their jobs. Common sense means I can recognize safety issues at their incipient state and once again, keep my crew safe."

See the difference? That's making the connection. The candidate drew a connection between himself, the job, and the department.

Always answer the question directly ("First, I am excellent with people"), then elaborate to make the connection ("As you know a fire captain must").

To help you make the connection and elaborate, think of yourself in your mind doing the job and simply describe to the panel what you see. Paint the picture in your mind of being the officer, describing how you got there, and what you are thinking. Let's look at another example.

The panel asks, "What would you do to improve operations in Dream Fire Department?" A simple answer would be, "I would set up a standard operating policy program."

Making the connection would say, "First, I would improve operations by assessing and improving the performance of my crew. We would train together to improve our weaknesses and build upon our strengths. That is my first priority. Ensuring my crew is operationally safe and effective. Second, I would enhance our current standard operating policies. By combining input from my crew with my own experience as an officer, I would create sample SOPs that target the key areas needing improvement in Dream Fire. Currently, those key issues are first-in engine company operations and RIC operations. I would pass those up to my BC when completed."

See the connection between the candidate, the job, and the organization? Keep those three components in mind to make the connection and give your answers depth and credibility. The panel will envision you in the job, and you will fulfill your goals (see fig. 9–8).

One final example would be the question, "What have you done to prepare for this job?" A simple answer would be, "I have taken my officer classes and been part of the apparatus committee. I have also done some acting captain work with my crew. I have led some drills as well."

An answer that makes the connection would be, "I'm so glad you asked! First I have dedicated myself to training. I have completed my fire officer classes, which have given me an excellent level of knowledge for command, tactics, and strategy. Next, that knowledge has then been put to the test as an acting captain where I have functioned successfully on eight working fires. Our crew was first due to three of them. I made key command decisions under pressure, and we successfully made aggressive interior attacks, minimized loss, and provided for safety. At no time did any of my crew get injured. On one fire, I was the incident commander and coordinated six companies including the tactical objectives of attack, ventilation, and rescue. Third, as part of my captain duties, I conducted drills on tactics, safety, and team work. The crew was very responsive and felt our training enhanced our effectiveness. Finally, as a member of the apparatus committee, I honed my team skills further by working alongside members of various divisions, met deadlines, and remained within budget. This has complimented my ability to train my crews as a captain, by understanding their needs and remaining on task. I will continue this commitment to train my crew as their company officer so that we all go home safely at the end of each shift."

Fig. 9–8 By making the connection between him and the job, he is helping the assessors imagine what it would look like with him in the job.

Whew, that guy is fired up! Did you notice that he transitioned from an actor to a full fledged captain as he painted the picture of his KSAs? That's making the connection!

Remember, making the connection is the art of connecting your past education and experience to the future of you doing the job. Elaborate and connect your skills to show that you are already in the mindset of the job you desire. The panel will imagine you doing the job as you describe a day in the life of your new position.

Interview — Score Sheet

For each dimension, rate the candidate from 1 (lowest score), to 5 (highest score). An total score of 70 is required for passing.

Candidate Name: ____________________

First impression (how does candidate set the pace for the interview):

1 2 3 4 5

Appearance (is candidate properly groomed, dressed, polished):

1 2 3 4 5

Attitude (does candidate have a healthy, positive attitude):

1 2 3 4 5

Sincerity (is candidate believable/honest):

1 2 3 4 5

Courtesy (is candidate respectful to board and position):

1 2 3 4 5

Self-confidence (is candidate confident without being cocky):

1 2 3 4 5

Enthusiasm/motivation (is candidate upbeat, positive, assertive, and a self-starter):

1 2 3 4 5

Understands questions/answers questions directly (does candidate grasp the questions and answer them directly and clearly):

1 2 3 4 5

Experience (does candidate have a good base to work from to be a safe and effective officer):

1 2 3 4 5

Education (has candidate sought further education):

1 2 3 4 5

Sociability/people person (is the candidate a team player, would he/she get along well with people):

1 2 3 4 5

Communication skills (does candidate listen and speak effectively and clearly):

1 2 3 4 5

Work ethic (does candidate believe in working hard):

1 2 3 4 5

Make the connection (does candidate let you know how he/she thinks and how his education/experience has prepared him/her, or does candidate simply recite his/her resume):

1 2 3 4 5

Fire department knowledge (does candidate have a good knowledge of the fire department's area, organizational hierarchy, priorities, culture, etc.):

1 2 3 4 5

Maturity (can candidate handle responsibilities of officer):

1 2 3 4 5

Calmness (is candidate relaxed, able to think methodically):

1 2 3 4 5

Preparation (is candidate prepared for the interview):

1 2 3 4 5

Strengths (does candidate have a clear grasp of his/her strengths):

1 2 3 4 5

Suitability (can the candidate do the job):

1 2 3 4 5

Total Score: ________

Comments:

Sample Questions to Consider

When reviewing these sample questions, do not attempt to memorize a scripted answer to each one. Simply have an outline in your mind from which you can draw upon during an interview.

1. Tell us about yourself.
2. Why do you want to be a lieutenant, captain, battalion chief, etc.?
3. What have you done to prepare for this job?
4. What are your short-, medium- and long-term goals?
5. What are your strengths/weaknesses?
6. What would you address with your crew in your first shift as captain/lieutenant?
7. What topics would you discuss with your officers in your first meeting as a battalion chief?
8. How would you improve our department's operations?
9. Where can we improve the department?
10. Where could we improve safety?
11. What is your leadership style?
12. What does integrity mean to you?
13. How would you handle sexual harassment?
14. How would you enhance customer service?
15. How do you instill trust and respect?
16. What type of officer will you be?
17. What would you do to motivate your crew(s)?
18. How would you implement the department culture and philosophy at your level?
19. What would you change about the department and why?

20. How would you handle a member with a drinking problem?

21. What if one of your firefighters was stealing?

22. How would you handle a chronically late firefighter?

23. What should the department do about succession planning for the future?

24. How would your style enhance the department's culture and leadership philosophy?

25. What could the department do about sick leave abuse?

26. How can the department streamline operations to remain within the budget?

27. What type of training is currently needed and why?

28. How would you implement that training? (If you made the connection, you would have already answered that in question #27)

29. How can we enhance our internal customer service?

30. What can we do to improve our risk management?

31. Do you have anything to add?

Index

A

B

C

D

E

F

G

H

I

J

K

L

M

N

O

P

R

S

T

U

V

W